30 Advances in Polymer Science

Fortschritte der Hochpolymeren-Forschung

With 113 Figures

Springer-Verlag
Berlin Heidelberg GmbH 1979

Editors

ISBN 978-3-662-15414-4 ISBN 978-3-540-35395-9 (eBook)
DOI 10.1007/978-3-540-35395-9

Library of Congress Catalog Card Number 61-642

Originally published by Springer-Verlag Berlin Heidelberg New York in 1979
Softcover reprint of the hardcover 1st edition 1979

2152/3140 – 543210

Contents

Polymer Analysis by Thermofractography

Egon Stahl and Volker Brüderle
Department of Pharmacognosy and Analytical Phytochemistry, University of the Saarland, D-6600 Saarbrücken, Germany

This survey informs on possible applications of thermofractography for polymer analysis. Its place within the known polymer analytical procedures is established and a detailed description of the apparatus required and the procedures are given. The essential characteristics of the new method are introduced in practical analytical examples and the predominant advantages underlined. Identifications of natural polymers, such as lignins, tannins, cellulose and proteins are described as well as those of plastics, such as phenol resins, epoxy resins and hardeners, polycondensates and vinyl polymers, also of plastic additives such as PVC plasticisers. In most cases, an identification scheme is given.

Table of Contents

I Introduction

Plastics based on organic polymers have found use in almost all branches of technology and construction as a result of their superior material properties. The desire to have available a suitable "material to measure" for every conceivable purpose has left its mark on their development. Natural and, predominantly nowadays, synthetic polymers, fulfil this requirement virtually ideally. Their physical properties result from a number of closely related and mutually influential attributes of the material and the preparation. The character of a macromolecular system is determined primarily by the chemical nature of the polymerisable monomer on which it is based. The synthetic route influences the external state and the uniformity of the polymer material. Macromolecules with sufficiently uniformly defined structure are obtained only when the relevant reaction parameters are strictly controlled; this is accomplished in nature through specific catalysts and in chemical technology through heeding the information about the process. The macromolecular properties of polymers depend also on modifying additives. These include mainly the secondary components of plastics, built into the polymer skeleton through copolymerisation or polyaddition, for example so-called hardeners of epoxide resins or copolymers of PVC with other polyolefines, also termed internal plasticisers. Numerous additives of low molecular weight, such as plasticisers, stabilisers, antistatic agents, play similar parts in determining the external nature of high molecular weight natural and synthetic products. Finally, the form of polymer raw materials is affected also by added fillers and pigments and by some processing aids.

The combined effect of all these factors on the structure of the polymer accounts for the multiplicity and wide variation in the external form and desired properties of organic polymers. In order to be able to form an opinion about the real structure, one must therefore acquire as much physical and chemical structural information as possible with the help of a comprehensive analytical scheme[1)].

These data are fitted together like in a mosaic to yield an image of the constituents and of the macromolecular structure based on these polymer even though this picture is never complete and is always capable of extension. However, this touches on only one set of questions concerning polymer analysis, although of central importance. Polymer analysis assumes economic significance in connection with the problem of industrial synthesis and production of semi- and fully synthetic plastics. Of special interest in this service branch of technological chemistry is information about the molecular and macromolecular structure of the products concerned, perhaps the formulation of patent requirements. Further, constant testing of the quality and uniformity, within set limits of tolerance, of macromolecular products of synthesis is indispensable. This analytical monitoring of production batches through tests of identity is aimed primarily at excluding possible undesirable side reactions. More comprehensive analytical problems concern the polymer composition; intermediate synthesis stages leading to the wrong products; and residual monomers in the end product. Knowledge of the impurities, additives and auxiliaries is also of considerable interest. Finally, identification of the structural units of the polymers enables conclusions to be drawn about the origin and purity of the raw materials employed.

Another analytical domain is the testing of competing products. The questions of detail here demand, however, an analytical outlay which increases linearly to exponentially with the profundity of information and accuracy demanded. Much the same applies to the answers to analytical problems of forensic chemistry, where, because of their uniquity, polymers are often an object of investigation[2].

In recent times the increasing health awareness of the public has given polymers analysis an established place in toxicological examination of packing material for medicaments and foodstuffs[3,4]. This brief outline of the aims and fields of application of polymer analysis gives a clear idea of the extensive and tricky tasks and questions of this analytical sphere. The fundamentals and informative value of the analytical techniques hitherto known and used for this are discussed in the next chapter. Their pros and cons are examined critically. This then leads to *thermofractography*, the sole procedure at present for fast thermal analysis of high polymers using *gradient degradation* coupled with thin-layer chromatography.

II Methods of Polymer Analysis

II.1 Preliminary Remark

Polymers can rarely be identified using the known methods for chemical determination of compounds of low molecular weight. This is due to fundamentally different structure of macromolecular substances and the resulting special physical and chemical properties[1]. As a rule, polymers are scarcely or only slightly soluble in the usual solvents; they can be vapourised generally only under decomposition; their molecular weights are by definition not uniform. The chromatographic and spectroscopic procedures which have shown their value when applied to compounds of low molecular weight furnish therefore results of only limited value.

II.2 Analytical Procedures Preserving the Polymer Structure

II.2.1 Spectroscopic Procedures

As expected, IR-spectroscopy[1,5–10], as a direct method, relatively simple to operate and non-destructive of sample, was long prominent in the development of a systematic scheme of polymer analysis for recognising characteristic structural elements of macromolecules. Nevertheless it became evident in the course of its application to nearly all problems of polymer analysis[5,7,9] that copolymers and also the various additives, impurities and auxiliaries of polymers could be identified only incompletely, especially when present in low concentration. The characteristic IR-bands of the various absorbing polymer segments are often superimposed on one another and on those of additives in the same wave length region; or the band intensities are too weak to permit an unambiguous assignment. Widespread application of UV-[1,5,10]

and NMR-spectroscopy[1,5,7,10] are subject to similar difficulties. Both methods possess a certain significance in the qualitative and quantitative determination of aromatic and conjugated unsaturated structures; and in ascertaining the crystallinity, tacticity, conformation, sequence length and degree of branching of elastomers. But for the polymer systems used in practice – mostly fairly complex mixtures – they provide information often of little value. High polymers cannot be submitted directly (if at all) to mass spectrometry, the most modern and sensitive spectroscopic identification procedure. As a result of their spatial structure and very high average molecular weights they are not vaporisable.

II.2.2 Separation Procedures for Fractionating Polymers

Along with the application of spectroscopic methods for direct identification of macromolecules suitable techniques have been developed for separation of oligomer fractions, components of polymerisates and additives to plastics. Fractionation on the basis of solubility differences[1,11] were those principally considered (fractional precipitation, extraction, gradient elution, counter-current distribution, ultra-filtration, dialysis, permeation, diffusion, sedimentation)[1]. Classical column chromatography (CC) steadily lost its importance in this sphere[1] and has been replaced largely today by high performance liquid chromatography (HPLC)[12–14]. Several effective separation procedures are available nowadays for separating macromolecules on the basis of their molecular size and molecular weight distribution. These differ markedly from one another in their outlay of apparatus. So-called gel permeation chromatography (GPC)[1] enables, for example, the relative molecular masses of various oligomer fractions to be determined; it can thus be applied to check the chemical homogeneity of polymers. High-voltage electrophoresis is still valuable, especially in polypeptide chemistry, for separating and purifying proteins according to their chemical nature and molecule size. Recently, both these established separation techniques have been supplemented to advantage by two further developments of thin-layer chromatography (TLC), a method really for separation of lipophilic substances of low molecular weight[15]. For example, according to experience so far reported, the newly introduced "Phase separation- or precipitation-TLC"[13] is highly suitable for characterising synthetic polymers according to their content of oligomer, homogeneity and degree of branching. Thin-layer electrophoresis prevails as a separation method in the domain of natural, mostly biogenic polymers[15].

II.2.3 Simple Physical and Chemical Test Procedures

The separation techniques mentioned above can be rationally employed only when suitable preliminary tests have established at least roughly the nature of the polymer and, in favourable cases, even the class of polymer. Such tests, especially important in large scale industrial chemical practice, include simple qualitative chemical tests for particular heteroelements and functional groups[1–6,10,11,17–26]. A certain number of physical data, mainly of interest for quality and production control, can also yield criteria for identification of an unknown polymer[1–6,10–11,17–26]. These include tests of mechanical and rheological properties, such as hardness, density, solubility, softening- and/or melting regions, viscosity, rigidity, ductility and dimensional stability. Other standard methods serve for acquiring optical (refractive index, colour, transmittancy), electrical (dielectric constant, specific resistance) and thermal (specific heat, coefficient of expansion, thermal conductivity and, last but not least, behaviour during thermal degradation) data. Their analytical usefulness is derived from comparison with tabulated standard values within definite limits.

During application of all the foregoing analytical techniques, the original polymer structure remains more or less unchanged. In the more physical methods of investigation the properties and activities of macromolecules can then be no more than summed values because the chemical factors influencing the individual structural elements of polymers, responsible for the macroscopic physical parameters, are externally additive. Consequently, conclusions about their presence and form, based on the data thus obtained, are possible only with reservation.

II.3 Analytical Procedures Involving Change of the Polymer Structure

Very useful hints about the analytical identity of polymer structures can be obtained from their changes and for those of physical reference values as a result of chemical reactions. For this, there is a number of methods, mainly physical, for degradating the primary polymer structure[27]. The desired analytical standard is attained, however, only by the thermal and hydrolytic degradation procedures when one looks at it from a practical and economic point of view. The thermal procedures in particular have proved to be especially suitable for depolymerisation of compounds of high molecular weight to defined fragments of low molecular weight. Moreover it is usually possible to combine them with suitable chemical or physical techniques of identification. Classical hydrolytic methods, especially for breakdown of condensation and addition polymers, have maintained their role in the analysis of polymers.

II.3.1 Differential Thermoanalytical Procedures

The techniques which are termed differential thermoanalytical methods[28–39](thermogravimetry, TGA; differential thermal analysis, DTA; differential scanning calorimetry, DSC; thermaloptical analysis, TOA; and dilatometry) combine in principle thermal degradation with a physical detector system; this enables one to follow the changes in substance-specific values brought about by the supply of thermal energy. They have the advantage of permitting the behaviour of polymers during degradation to be followed quantitatively in a temperature gradient. Their shortcoming, the same as with the simpler physical test procedures, is that individual polymer segments, components or additives cannot be recognised. On the contrary, these techniques record no more than overall losses of mass, volume changes and alterations in optical properties resulting from elimination of secondary products, mostly of low molecular weight and gaseous.

II.3.2 Thermal Degradation in Combination with Chemical Separation and Identification Techniques

In order to characterise the individual products derived from thermal degradation of polymers, suitable separation and detection systems were sought, if possible directly coupled with the degradation unit. It is a fact that the type and number of possible thermolysis products is characteristic for the polymer composition and permits conclusions to be made about the structure of the polymer. A decisive breakthrough in the domain of chemical polymer analysis was accomplished by the development of thermoanalytical degradation procedures in indirect and direct combination with physico-chemical separation and detection methods.

II.3.2.1 Combination of Pyrolysis and IR-spectroscopy

IR-spectroscopy was employed at first for identifying the pyrolysis products of polymers[5, 8, 40–44]; considerable difficulties in interpretation were encountered, however, where no prior separation of the products was carried out.

II.3.2.2 Pyrolysis-GC and Pyrolysis-MS and Coupling of Both Procedures

The methods of pyrolysis-gaschromatography (PGC)[43–74] and of pyrolysis-mass-spectrometry (PMS)[75–82], both based on isothermal degradation, have proved to be highly efficient. They are distinguished by high sensitivity and outstanding reproducibility. They provide the possibility of compiling tables of "finger print" pyrograms for every class of polymer[5]. The combination of both procedures, pyrolysis-GC-MS, in the so-called "on-line" technique[56, 61, 62, 81], furnishes even more informative analytical results. This analytical system while admittedly demanding a considerable outlay of apparatus, enables degradation, separation and identification to be carried out smoothly in a single working stage. Further, these pyrolysis procedures can be used for quantitative analysis of the polymer composition. However, despite these virtues, they have only limited value in the routine analysis of polymers.

II.3.2.3 Limits of Polymer Analysis by Pyrolysis-GC and -MS

1) Restriction to isothermal conditions.
2) Thermal breakdown processes cannot be followed.
3) No distinction of the polymer additives, residual monomers and thermolytically produced degradation products, all of which arise in differing temperature regions.
4) Polymer components cannot be pre-fractionated chemically during the thermal degradation.
5) Large demand of time and sample (> 10 mg).
6) Considerable requirement of money and personnel for equipping and manning the analytical system.

II.3.3 Thermal Degradation of Polymers in Combination with TLC

There has hence been no lack of attempts to introduce into polymer analysis simpler, faster and less costly separation and detection procedures for the products of thermolysis of polymers, if possible directly coupled with the thermolysis unit. Thin layer chromatography (TLC) was a first candidate here[15]. It is superior to more complex apparatus as far as ease of operation, economy and duration of analysis are concerned; and its sample requirement, efficiency of separation and reproducibility of analytical data are at least as favourable. In addition, many detectors, specific for groups and for individual substances[15,84] are already available in TLC; and several samples can be analysed and compared alongside one another[15,85]. These advantages stimulated the study also by TLC of the products of pyrolysis of polymers[1,5,6,25,86–93].

Notably Braun and co-workers have developed a useful aid for polymer analysis, involving relatively crude pyrolysis in a test tube and subsequent manual transfer of the pyrolysate to TLC; they give a scheme for separation of plastics based on this[25,26,89–92].

II.3.4 Hydrolytic Degradation Procedures in Combination with TLC

Classical alkaline and/or acid hydrolysis is often employed for defined break-down of hydrolysable polymers (condensation and addition polymers). The fission of the macromolecules then usually reverses the reaction of formation. Moreover, preliminary chemical separation of the hydrolysis products into basic, acidic and neutral parts is facilitated in the basic or acidic medium. Hydrolytic degradation of such polymers is thus combined with a chemical pre-fractionation of the components of hydrolysis. Almost only TLC has been used for separation and detection of the hydrolysis products[94–102]. Only occasionally has GC found use for this[101,103]. Disadvantages are, however, the extremely long time of hydrolysis (sometimes > 18 h) and the relatively large sample (1–2 g). More recently, the hydrolytic breakdown of condensation polymers has been extended and accelerated by introducing the old established method of alkali fusion, then followed by GC-identification of the hydrolysis products[104]. Problems arise, however, of the "point application" and of interference by larger amounts of water during the chromatography.

II.3.5 Coupled Procedure Thermolysis-TLC

II.3.5.1 TAS-Procedure

The TAS-procedure, developed originally for rapid extraction of the drug content of natural plant material, was an important move towards a thermolysis-TLC-coupling[105–108]. During the initial testing stage of the new coupling procedure, the possibility of characterising polymers through TLC of their products of thermal degradation was pointed out[108]. Even before this, it had been shown that an experimental arrangement, essentially that of the TAS-procedure, with direct combination of pyrolysis unit and TLC-separating system, was suitable for polymer analysis[109].

II.3.5.2 Thermofractography

The further development of the TAS-apparatus led then to thermofractography (TFG)[110]. The decisive new feature is the temperature-programmed degradation of polymers using a pre-selectable linear increase of temperature, accompanied by a simultaneous movement of the thin layer which is synchronised with the temperature rise. Apart from the more physical breakdown procedures of differential thermal analysis, thermofractography offers for the first time an analytical technique for continuous degradation of macromolecular substances in a temperature gradient and coupled directly with TLC[111]. The individual chemical characterisation of polymer fragments through their "thermofractogram" provides a qualitative and semi-quantitative identification of natural and synthetic polymers. The procedure is by no means limited to thermal degradation. The construction and instrumental side of the TFG-apparatus (TASOMAT) can be sufficiently modified so as, for example, to be able to carryout even the alkali fusion in it as a hydrolytic procedure for condensation polymers.

Both coupled techniques are described below and the various procedural variants are discussed. The equipments expressly designed for these purposes are presented. In the main part of this survey, the application and performance of TFG in the analysis of natural and synthetic polymers are then shown. Also below is a compilation of the most important methods for qualitative polymer analysis (see Scheme 1).

III Presentation of the Method

III.1 TAS-Procedure and Thermofractography

The TAS-procedure was described in 1967–68 as a simple thermal micro-separation and application procedure coupled directly with thin-layer chromatography[105–108]. Substances which are volatile at higher temperature are hereby separated from involatile material and brought directly on to a thin-layer chromatographic start point. For this, a few mg of sample are introduced into a special glass cartridge, one end of which is drawn out into a capillary; the other end of this so-called TAS-cartridge is closed with a silicone membrane. The cartridge is pushed into the TAS-oven maintained at a definite temperature. Within 1–2 min the volatile substances leave the capillary exit as a fine stream of vapour and condense as a start point on the thin-layer plate held directly opposite the exit (Fig. 1).

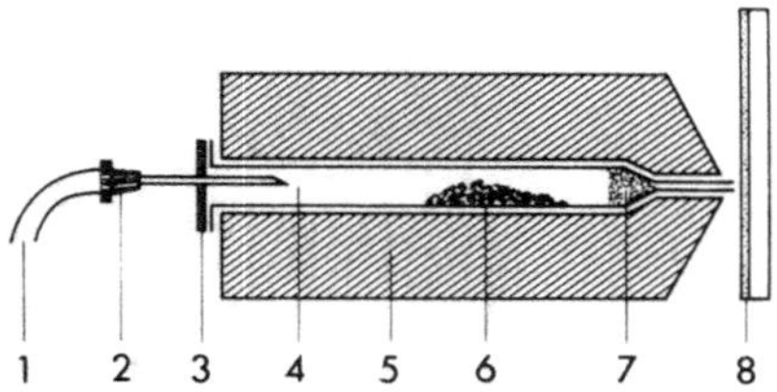

Fig. 1. TAS oven, longitudinal section: 1 carrier gas, 2 injection needle, 3 seal, 4 TAS glass cartridge, 5 heating block, 6 sample, 7 quartz wool, 8 TLC plate with layer

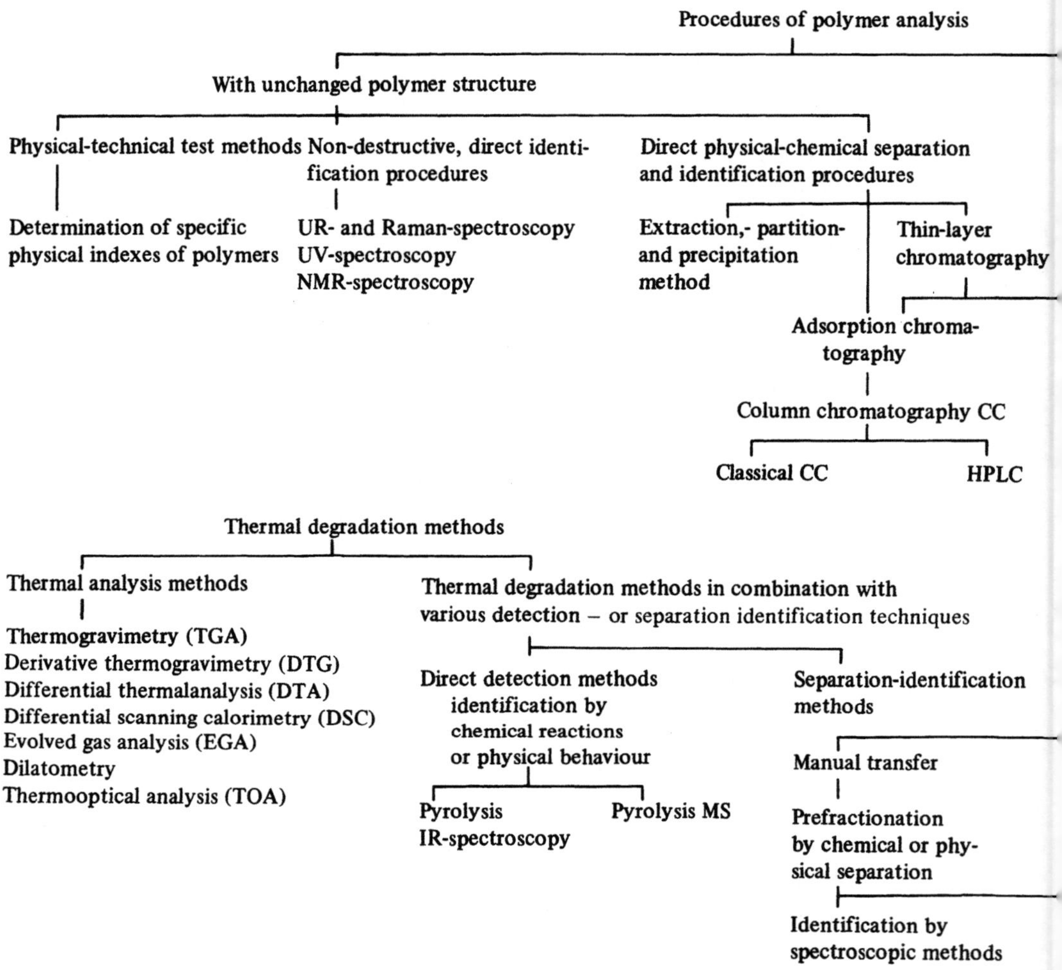

Scheme 1: Analytical methods in the field of polymers

- With changed polymer structure
  - Chemical pretests
    - Rapid test method on hetero elements and functional groups
      - Determination of chemical indexes
        - Acid number
        - Saponification number
        - Double bond equivalents
    - Elementary analysis
  - Thermal degradation procedures
  - Hydrolytic procedures

- Phase separation or precipitation TLC

Hydrolytic procedures
- With manual transfer to detection system
  - See thermal degradation methods
- On-line techniques
  - Isothermal
    - Pyrolysis-GLC
    - TAS-method
  - Gradient degradation
    - Thermofractography of polymers, alkali fusion

- On-line techniques
  - Pyrolysis-GLC
  - Pyrolysis-GLC-MS
  - TAS-method and thermofractography (TFG)
- Chromatographic methods
  - PC
  - TLC
  - GLC
  - HPLC
  - Identification by chemical or spectroscopic methods

The effectiveness of transport of the sample to the thin layer can vary. "Forced convection" gives the smallest yields. Appreciably better results are attained when a so-called propellant, that produces steam for example, is added. Best of all, however, is to carry out a carrier gas distillation by passing a current of inert gas at low speed over the sample, *i.e.*, through the TAS-cartridge. The TAS-procedure is isothermal and makes use of a stationary thin-layer. In the thermofractographic method, however, the sample is heated gradually from room temperature to a maximum temperature of 450 °C. At the same time, the thin layer is slowly moved forward, thereby yielding a band of condensed substances fractionated according to their volatility. This coupling of linear temperature increase of the sample with a fractionated capture of the volatile or pyrolysed substances is termed thermofractography (TFG)[110]. A special apparatus, called the TASOMAT, is used to carry out the procedure. It can be employed for the TAS-procedure, for preparative winning through band condensation under isothermal conditions, and also for the procedure, just described, of fractional collection of the substances which have been heated in the temperature gradient. The result of the described procedure after chromatography is termed a thermofractogram. The substances are separated along the abscissa of the thin layer according to their boiling or sublimation temperatures or thermolysis behaviour; and along the ordinate according to their chromatographic behaviour.

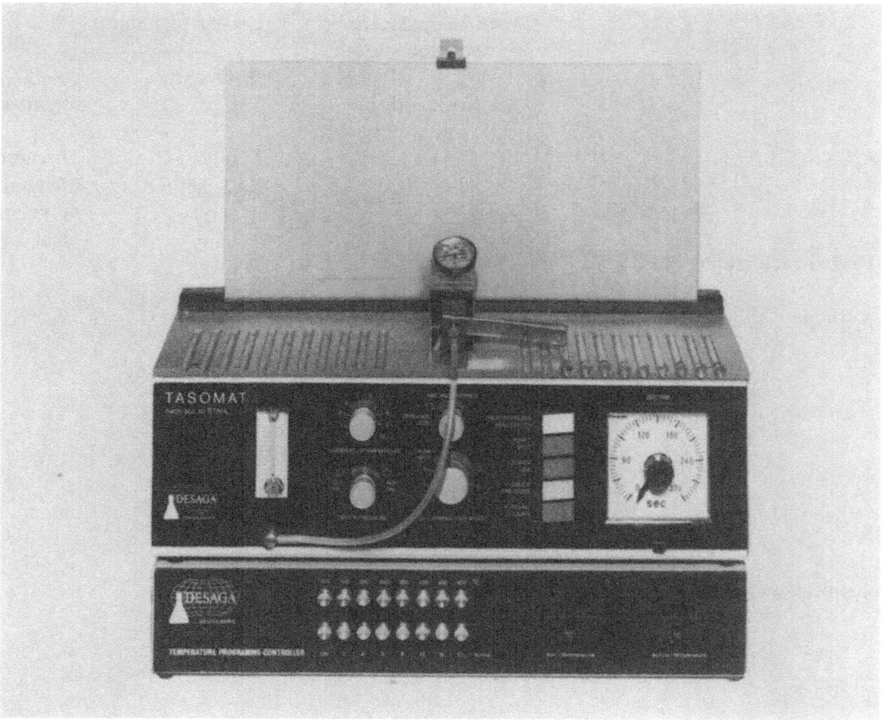

Fig. 2. Front view of the *Tasomat* (above) and control unit (below)

III.2 Apparatus

The TFG-apparatus is known under the name of TASOMAT (Fig. 2, above). The oven block (Fig. 1) has the same dimensions as the TAS oven block but is made from a steel alloy, of lower coefficient of thermal expansion than aluminium. This is important for maintaining constant the small gap between cartridge tip and collecting layer as the temperature rises.

Heating to the maximum temperature of 450 °C is accomplished by a heating cartridge built into the block. For temperature setting under isothermal conditions (TAS-procedure) the apparatus has a built-in thyristor stepless control device. However, in order to obtain a linear rise in temperature of the oven block and various rates of heating, a specially adapted control unit must be available (Fig. 2, below). The desired rate of heating and end-temperature can be selected with a press button device on the control apparatus. The actual and set temperature values of the oven block are given digitally. At the end of an experiment, the oven block can be cooled rapidly with a built-in ventilator. The gap between the tip of the TAS-cartridge and the thin layer can be rapidly regulated by turning a screw knob. The size of the gap is seen from the projected shadow of the cartridge tip on the thin-layer (Fig. 3).

The TAS-cartridge is made of Pyrex glass and of such length and thickness that it fits tightly in the partitions of the oven block. In contrast to the usual procedure, a sheath made of aluminium foil, open at both ends, is inserted into the TAS-car-

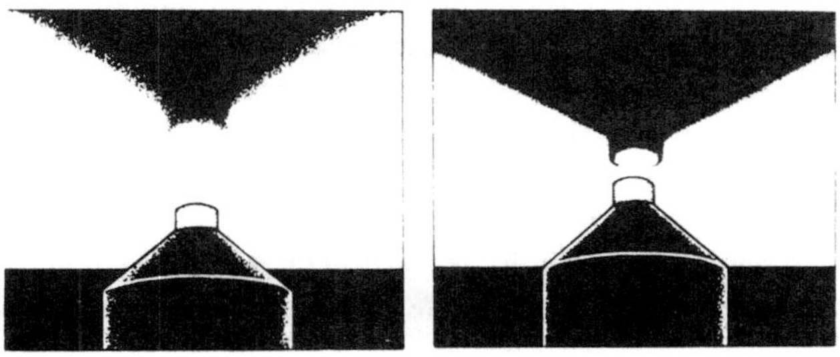

Fig. 3. Distance regulation by shadow projection of the oven tip on to the TLC plate. Left: distance too far, blurred projection; right: distance under 1 mm, exact projection on to the thin layer

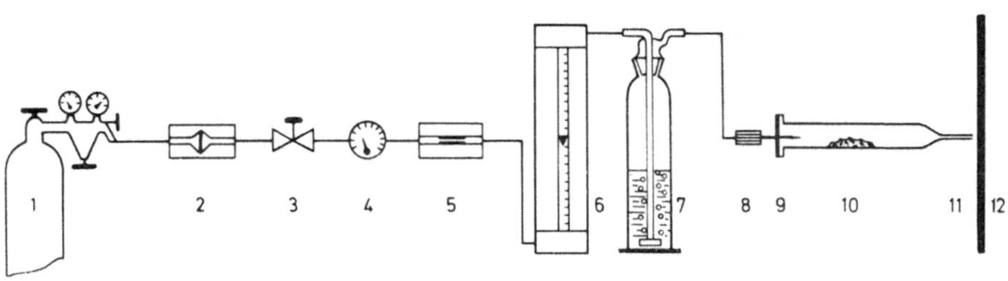

Fig. 4. Flow diagram, from the nitrogen cylinder (1) to the TLC plate (12): 2 filter, 3 fine adjustment valve, 4 manometer, 5 reducing capillary, 6 flow meter, 7 special purposes washing bottle for the saturation of gas, 8 injection needle, 9 seal, 10 TAS cartridge with sample, 11 capillary of the TAS cartridge

tridge. The sheath is loosely stoppered at one end with glass wool and after inserting the sample, also at the other end. The sheath is then put into the TAS-cartridge which is tightly closed with a so-called HD-clip. The current of inert gas is passed through the membrane seal by means of a very fine, short stainless steel injection needle (18 x 11). Usually pure nitrogen is used or, better, helium[112].

The gas stream is controlled by a fine pressure regulator in combination with a flow meter. An appropriate wash bottle can be inserted into the stream in order to saturate the gas, if desired, with, for example, water vapour or a solvent. The flow chart of such an assembly is shown in Fig. 4.

TLC-Plate Transport

The TLC-plate can be transported at 10 different speeds, ranging from very slow to fast. Both the normal 20 x 20 cm TLC-plates and 20 x 40 cm plates can be placed vertically on the plate carriage. Both the application length and position, *i.e.*, the lateral position of the application band on the layer, can be regulated. The Japanese version of the TASOMAT has a number of constructive alterations but it functions as described above. The TASOMAT with control unit is basically intended for three different procedures:

a) Spot Application of Individual Samples (Analogous to the TAS-Procedure). Experience up to now with the TAS-oven in routine work has shown that one sometimes forgets to transport the TLC-plate further by hand before a new sample is introduced and also fails to adhere to the heating time. In order to prevent both these things, a micro-switch has been built into the TASOMAT which, as soon as the TAS-cartridge is pushed in, switches on an electric control clock with a setting range from 0 to 300 sec. After the pre-selected time has run, a humming noise persists and stops automatically when the TAS-cartridge is removed. At the same time the transporting motor receives an impulse and displaces the TLC-plate 1.5 cm to the next place for application. In this way, double application or exceeding the time cannot happen.

b) Vapour Deposition as a Band. Vapour deposition as a band is a good way of obtaining volatile substances from samples on the micro-preparative scale. This requires a rapid TLC-plate movement (280 cm/min) and automatic reversing of the direction of run when the chosen length of band is attained.

c) Linear Heating with Fractional Capture as a Band. In thermofractography the rate of heating has first to be chosen and then correlated with the rate of plate movement and the length of the TLC plate. A scale of temperature for the chosen conditions is set up and can be later used for interpretation of the thermofractogram. The customary standard conditions apply to the TLC. It is an advantage here to run a reference mixture at the same time and to work out all separations in detail beforehand.

III.3 Execution and Interpretation

a) Filling the TAS-Cartridges: Usually from 1 to 10 mg of sample is taken, established in preliminary tests. Excessive amounts giving too large TFG-zones are to be avoided. The sample is in the sheath of aluminium foil which should be about in the middle of the TAS-cartridge.

b) Regulation of the Current of Carrier Gas: After the TAS-cartridge has been filled and closed with the HD-clip and introduced into the cold oven block, the sealing membrane of the HD-clip is pierced with the injection needle. The flow of gas is adjusted previously to 10–20 ml/min by means of the fine regulation valve and the flow meter (ball position 0.7–1.3). The flow meter is calibrated with the help of a 10 ml soap bubble flow meter when the apparatus is first used.

c) Choice of TLC-Plate and Mounting: Generally, the standard glass plates of 20 x 20 cm are used. If longer application bands are desired, two of these plates or a 20 x 40 plate can be mounted in the groove of the plate carriage. The layers are customarily of silica gel. No special precautions against evaporation during application are required during the occasional use of impregnated silica gel layers.

d) Start Position, Length of Application Band and Adjustment of the Gap Between Cartridge Tip and Layer: The plate carriage is pushed by hand (knob "Transport" in the intermediate position) so that the extremity of the oven, *i.e.*, the TAS-cartridge, is about 1 cm from the left side of the TLC-plate. An adjustable screw, running on a rail behind the carriage is then tightened firmly. The same thing is carried out with the adjustable screw on the right side. The length of the application band on a normal 20 cm plate is now 18 cm. The gap between cartridge tip and layer is adjusted to about 0.5 mm by means of the knob "distance" with the assistance of the shadow projection. The setting of the gap is checked about every 100 °C during heating and adjusted if necessary.

e) Choice of Rates of Plate Transport and Heating: The speed of plate transport can be increased stepwise from 0.28 mm/min. The linear increase of temperature is operated by pressing the appropriate knob on the control unit (Model Series II, with digital readout since 1976; from this date the transport speeds have been slightly altered as a result of changes in the TASOMAT gears). The available heating rates are 2, 4, 6, 8, 12 and 16 °C/min. The end temperature can be set also through pressing a button at values ranging from 100 °C to 450 °C in steps of 50°. The "set-temperature" value runs at the selected heating rate to the desired end value. The corresponding oven temperature, "actual temperature", is also shown digitally alongside. For our experiments on thermolysis of plastics we preferred the settings in Table 1, depending on the problem:

Table 1.

Rate of heating in °C/min	Rate of plate transport in mm/min	Length of run from 150–450 °C in cm	Temperature scale in mm/°C
4	2.81 (Stage 1)	21	0.17
8	2.81 (Stage 1)	10.5	0.35
12	5.60 (Stage 2)	13	0.466
16	5.60 (Stage 2)	10.5	0.35

It must be mentioned that analyses began often at 150 °C, with end temperature of 450 °C, after which it continued isothermally for a further 5 cm.

f) Scale of Temperature for the Thermofractogram: A scale of temperature which corresponds to the working conditions is needed for determining the zones in the abscissa direction in the interpretation of a thermofractogram. A correlation between the position of application and the oven temperature is derived from the rate of heating and the transport of the TLC-plate. If the set rate of plate transport is divided by the rate of temperature increase, the stretch is obtained through which the TLC-plate is moved while the oven temperature rises 1 °C. The values for the given example are quoted in column 4 of the table. A scale is based on these, with intervals of 10 °C. The scale can be placed on the start point of the TFG and enables the temperatures of the beginning and end of the zones to be read off.

g) Chromatography and Detection: The thin-layer chromatography is carried out under standard conditions. A reference mixture of dyes is always co-chromatographed on the right or left of the layer in the TFG of unknown substances. Where, for example, preliminary tests have furnished information about the nature of the polymer samples, as many as possible of the expected fragmentation products are applied for reference. In this case, the best mobile phase can also be ascertained in prior experiments. Visualisation of the zones is first tried by using layers containing a fluorescent indicator and inspecting them in short- and long-wave UV light. Subsequently, so-called universal spray reagents, such as anisaldehyde-sulphuric acid, followed by heating, are used. Many other possibilities are described in the appropriate articles and standard works on TLC.

h) Interpretation of the Thermofractogram: The thermofractogram is interpreted in two dimensions. In the direction of the ordinate the Rf-values of the various zones are obtained, which inform on the "polarity" of the transfer products. The temperature at which the zone begins is then derived in the direction of the abscissa. The length and width of the zone depend on the amount of substance which reaches the layer; this yields an idea of the quantitative composition. The shape of the zone is also noted. The most commonly occurring form is the simple spindle shape. Also encountered are double spindles, so called A-clubs with the wide end at the beginning (A from Anfang, German for beginning), and E-clubs with the fatter part at the end. Sickle or ribbon-shaped zones can occur also. The best type of description is to draw the thermofractogram schematically (cf. for example, Fig. 6). Unknown zones can be identified by IR- or UV-spectroscopy, gas chromatography or mass spectrometry, after transfer.

III.4 Procedure Variants in Thermofractography

a) Carrying out Chemical Reactions in the TAS-Cartridge. There are numerous chemical reactions which take place only at higher temperatures. TFG offers the chance of carrying out such reactions on the micro-scale. The volatile reaction products which are formed can then be identified on the chromatogram.

The factors governing classical distillation with zinc dust were first investigated and examples quoted[115]. Zinc dust activated with copper proved to be the best.

From 5 to 100 μg of substance in solution is added to the zinc, the solvent removed and the mixture put into a TAS-cartridge according to the scheme given in Fig. 5 A. The loaded cartridge is then introduced into the heating block and after leaving for 30–45 sec, nitrogen as carrier gas is passed through at 15 ml/min. There is the choice of working isothermally with spot collection on the thin layer; or in a temperature gradient (*e.g.* 200 to 450 °C) with simultaneous transport of the TLC-plate.

The classical sulphur or selenium dehydrogenation can also be performed on the microgram scale in the TAS-cartridge[116]. The use of a specially prepared dehydration-dehydrogenation mixture (potassium hydrogen sulphate-sulphur or selenium) has proved advantageous. The volume of the TAS-cartridge (2 ml) must be reduced to 0.2 ml by means of a stainless steel cylinder (Fig. 5 B). Depending on the type of substance, the reaction temperature lies between 160 and 300 °C. Catalytic dehydrogenation with palladium-barium sulphate has proved to be very good[117]. The reaction takes place according to principles of dehydration in the gas phase through injection of the investigated solution in batch amounts into the TAS-cartridge filled with cata-

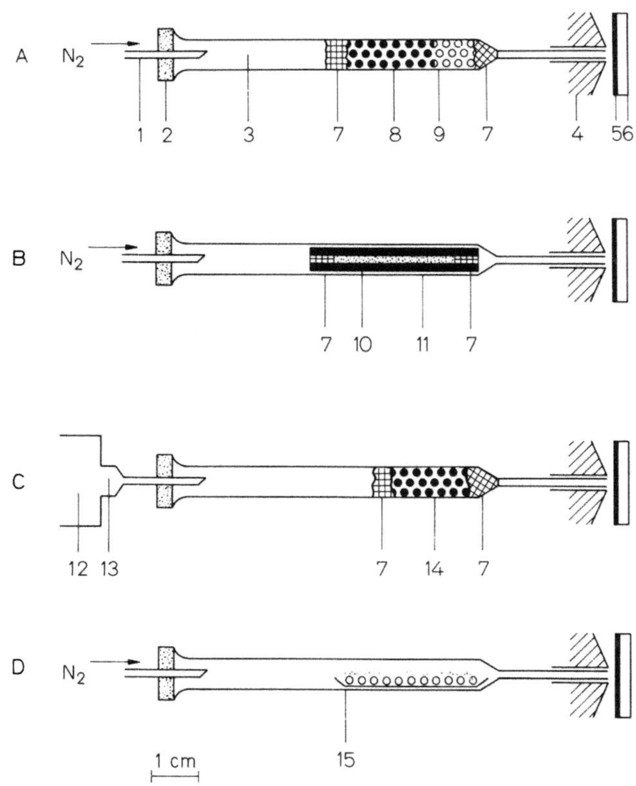

Fig. 5. Different filling possibilities of the TAS cartridge: A for zinc dust distillation, B for sulphur and selen dehydrogenation, C for catalytic dehydrogenation, D alkali fusion in a nickel boat. 1 injection needle, 2 seal (silicone membrane), 3 glass cartridge, 4 tip of oven block, 5 thin-layer, 6 carrier plate, 7 quartz wool, 8 zinc dust with substance, 9 zinc dust without substance, 10 dehydration mixture with substance, 11 stainless steel cylinder, 12 micro doser according to Dibbern, 13 micro syringe, 14 catalyzer, 15 reaction boat

lyst (Fig. 5C). Sample sizes are of the order of 20--100 μg and reaction temperatures are between 200 and 400 °C, depending on the compound. The reaction products are transferred to the layer by the solvent vapour.

b) Alkali Fusion in the TAS-Cartridge[118]. Alkali fusion is a method for decomposing (alkali hydrolysis) organic and inorganic compounds which has likewise been known for a long time. Only recently has it been used successfully in the analysis of polycondensed compounds, however. The fusion mixture is held in a nickel boat which fits into the TAS-cartridge; it consists of 3–5 mg of sample and ten times this mass of a mixture of potassium hydroxide and sodium acetate (mp. c. 220 °C). The filled boat is pushed into the TAS-cartridge, which is then inserted into the heating block (Fig. 5D). Prior to this, nitrogen is passed through for 1 min. The cartridge is left in the oven block for 3 min, during which nitrogen is led through at 15 ml/min. The volatile products formed, *e.g.*, polyamines, polyols, collect on the thin layer and are subjected to chromatography. After the boat has cooled down, its contents can be neutralised with 85% ortho phosphoric acid. Volatile organic acids can then be transferred to another TLC-plate by reheating, and identified by chromatography.

IV Thermofractography of Natural Polymers

IV.1 Lignins[119, 120]

Introduction: Lignins are usually classified into three groups, namely (a) softwood (gymnosperm) lignins; (b) hardwood (angiosperm) lignins, and (c) grass lignins. The differences between these groups arise from the proportions of guaiacyl-, syringyl-, and p-hydroxyphenylpropane structural units in the polymeric lignin molecules. Thus softwood lignins consist mainly of guaiacyl-propane units while hardwood lignins contain both guaiacyl- and syringyl-propane units. Grass lignins are presumed to be made up of all three, guaiacyl-, syringyl-, and p-hydroxyphenylpropane units.

On oxidation with nitrobenzene[121, 122] in an alkaline medium, softwood lignins give mainly vanillin, hardwood lignins give in addition syringaldehyde, and grass lignins give substantial amounts of p-hydroxybenzaldehyde besides these two aldehydes.

In this respect ethanolysis of lignins to the so called "Hibbert ketones" would appear to be a more valuable method[123]. This, however, has been used relatively infrequently, because of its greater difficulty and its low yields. More recently, Kratzl *et al.*[124] made use of the pyrolysis products of lignins as a tool for distinguishing between gymnosperm and angiosperm lignins. Connection of the pyrolysator to a vapor phase chromatographic unit showed that sulfate lignins from angiosperm give 4-methyl-2.6-dimethoxyphenol, which is absent in the gas chromatograms derived from gymnosperm lignins.

Chromatographic conditions: In order to optimise the conditions, a mixture of the expected thermolysis products was developed on silica gel GF_{254} under standard conditions with various solvents. The best results were obtained with a mixture of chloroform-ethylacetate (95 + 5). Before it is sprayed with a specific reagent, the plates should be examined under short wave UV-light (254 nm) in order to detect the substances by means of fluorescence quenching. Fast blue salt B has proved to be the best staining reagent, giving a spectrum of different colors with most of the phenols of interest. The differences in color are improved by a further spraying with hydrochloric acid. With fast blue salt B colors are obtained which depend on the substituent in the 4-position of the phenol. Thus, for example, guaiacol derivatives give colors ranging from blue to red to violet, while syringyl derivatives stain blue to green. The specificity and sensitivity of this reagent are higher than those of the anisaldehyde-sulfuric acid reagent which we had previously used. Under these conditions, the compounds shown in Table 2 were chromatographically analysed and their hRf values and color reactions estimated.

Besides the lignin thermolysis products with a known structure, Table 2 contains nine dimeric lignin model compounds (6, 8, 9, 14–17, 20, 23) and some colored thermolysis products of unknown structure.

Thermofractography of lignins: Milled wood (MW) lignins, prepared according to Björkman[125)], were used from pine (Pinus silvestris), spruce (Picea excelsa), beech (Fagus silvatica), and poplar (Populus nigra). The MW-lignins from two European mistletoes (Viscum album), which were grown on pine and poplar, were also examined. – Two typical TFG's from spruce and beech lignin are shown in Figs. 6 and 7. From the TFG's it can be seen that the thermolysis products are released from lignin in two temperature regions, the first reaching from 225° to about 350 °C and the second from 350° to 450 °C. With gradually increasing temperature, coniferyl and sinapyl alcohols are the first substances to be released from lignin at about 300 °C, followed by the corresponding aldehydes at 325 °C, the alkyl derivatives between 350 and 450 °C, and finally the dealkylated phenols guaiacol and 2.6-dimethoxyphenol are released: at 375 to 425 °C. The results are summarised for comparison in Table 2.

Evaluation and Discussion of the Thermofractograms:

1) Softwood lignins (spruce and pine) give guaiacol, 4-vinylguaiacol, vanillin, coniferyl alcohol and coniferaldehyde, and also the violet coupling zone 22, and the yellow-to-orange substance 24. As these substances are also obtained from all other lignins, they are typical finger print substances for lignins in general.
2) Hardwood lignins (beech and poplar) give, besides those substances mentioned unter 1) 2.6-dimethoxyphenol, 4-vinyl-2.6-dimethoxyphenol, sinapaldehyde, and sinapyl alcohol, and also an orange (25) and a red colored substance (27) in Table 2.

At lower temperatures, between 225 and 350 °C, the first substances to be released are phenols with an unsaturated side chain (vinylguaiacol, coniferyl alcohol, coniferaldehyde, and the corresponding syringyl derivatives), while guaiacol, 2.6-dimethoxyphenol, and their saturated 4-alkylderivatives appear only at temperatures above 350 °C. This possibly indicates that the β-aryl and/or the benzyl ether bonds between the C_9-structural units are the first to be split giving unsaturated thermolysis products, and the carbon-carbon bonds between the side chains are split later, when the temperature reaches about 350 °C.

In model experiments with dilignols this assumption could be confirmed. On heating the guaiacylglycol-β-cresol ether (16), vinylguaiacol (1) was obtained as the main thermolysis product. Besides 1, guaiacol, vanillin, and some unknown sub-

Table 2. hRf-ranges and detections of model compounds and thermolysis products of lignins

Substances	hRf-range	Colour reaction with Fast Blue Salt B	Lignins of Beech	Poplar	Spruce	Pine
1) Vinylguaiacol	70–75	Violet blue	+	+	+	+
2) Guaiacol	69–74	Yellow red-violet	+	+	+	+
3) Furfural (from hemicelluloses)	68–70	–				
4) 4-Vinyl-2. 6-dimethoxy-phenol	64–68	Violet green	+	+	○	○
5) 4-Ethyl(or methyl)-2. 6-dimethoxy-phenol	59–64	Violet green-blue	+	+	○	○
6) 2.2′-Dihydroxy-3.3′-dimethoxy-5.5′-di-n-propyl-biphenyl	56–60	Brown				
7) Phenol	50–55	Yellow pink-violet				
8) Diisoeugenol	50–55	Brown				
9) Divanillyl	50–55	Brown				
10) Vanillin	48–50	Faint yellow	+	+	+	+
11) Syringaldehyde	45–48	Brownish	○	+	○	○
12) Coniferaldehyde	42–46	Orange	+	+	+	+
13) Sinapaldehyde	40–42	Olive green-blue	+	+	○	○
14) Disyringyl	36–40	Ochre				
15) 1.2-Bis-(4-hydroxy-3.5-dimethoxy-phenyl)-1-propanone	30–35	Ochre				
16) 1-(4-hydroxy-3-methoxy-phenyl)-2-(2-methoxy-4-methyl-phenoxy)-1-ethanol	16–20	Violet				
17) Pinoresinol	16–20	Brown				
18) Coniferylalcohol	16–20	Violet-blue	+	+	+	+
19) Sinapylalcohol	12–16	Violet blue-green	+	(+)	○	○
20) Syringaresinol	12–16	Brown				
21) Pyrocatechol	9–11	Orange-brown				
22) Zone 22	5–10	Violet	+	–	+	+
23) Dihydro-dehydrodiconiferylalcohol	0–5	Ochre				
Zones, visible before spraying						
24) Yellow-orange substance	48–50		+	+	+	+
25) Orange substance	45–48		+	+	○	○
26) Blue substance	44–46		+	○	○	○
27) Red substance	35–45		+	+	○	○

stances were obtained but only in small amounts. On the other hand, the seven dilignols 6, 8, 9, 14, 15, 17, and 20, listed in Table 2 which have a carbon-carbon bond between the aryl units, gave vanillin, guaiacol, 4-alkylguaiacols or the corresponding syringyl derivatives as the main thermolysis products, while phenols with an unsaturated side chain were absent or occurred only in minor amounts. The biphenyl compound 23 (cf. Table 2) was thermally remarkably stable; even when it was heated to 380 °C before the nitrogen flow was turned on, it passed over without decomposition.

From these results one can conclude that the bonds between the C_9-units in lignin exhibit different behaviour during thermolysis. Thus, the nature and yields of thermolysis products depend, at least to some extent, on structural features of the lignin macromolecule and give further details for distinguishing between lignins from different plants.

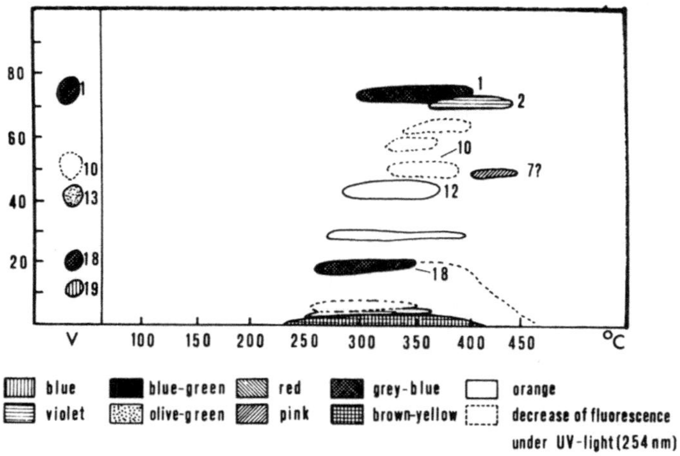

Fig. 6. Thermofractogram of 2 mg spruce lignin. The colours are obtained after spraying with Fast Blue Salt B. The numbers refer to the substances in Table 2. V = reference solution. For further details see[119–120)]

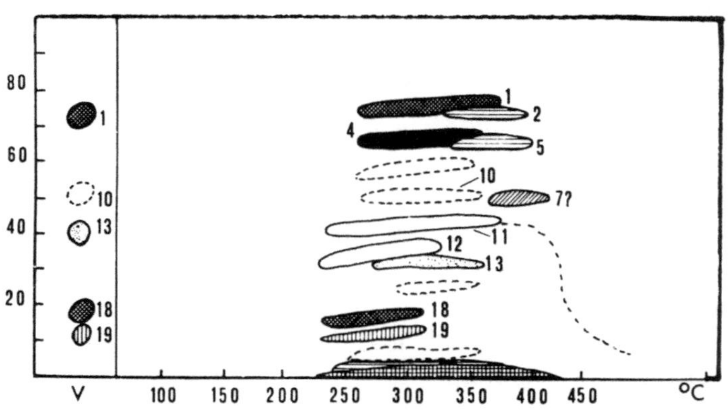

Fig. 7. Thermofractogram of beech lignin. For comparison see Fig. 6

The good discrimination between the guaiacyl and syringyl derivatives on the TFG's (cf. Fgs. 6 and 7) makes the classification of lignins into softwood and hardwood lignins extremely simple. Softwood lignins give guaiacyl derivatives, while hardwood lignins give in addition at least four syringyl derivatives. The thermofractographic method is of special interest for the classification of lignins other than softwood and hardwood lignins, for example lignins from herbaceous angiosperms, ferns and mistletoes.

The experiments yielded a quick answer to the question about which type of lignin is synthetised by mistletoes. It produces a hardwood lignin, even when growing on a softwood tree.

IV.2 Tannins and Leather[126)]

According to Freudenberg[127)] the naturally occurring tannins can be classified as follows:

a) Hydrolysable Tannins: The gallotannins and ellagitannins belong here. Their most important structural unit is gallic acid, which can occur as such, as m-digallic acid, hexahydroxydiphenic acid or ellagic acid, generally esterified with glucose. The various tannins of this class are yielded by the different ratios of gallic acid to glucose.

b) Condensed Tannins: These so-called flavanoid tannins are based on hydroxylated flavanols, notably catechin and epicatechin. Condensation of these leads first to the water-soluble catechin tannins and then, during storage, to the phlobaphenes of higher molecular weight and hence insoluble in water.

c) Mixed Tannins: Ester formation between the above mentioned primary units, gallic acid and catechin, gives the catechin gallates. Further combinations are conceivable.

Alongside these naturally occurring tannins there are synthetic organic substances which have gained significance as tanning agents. These are based on phenol or naphthalene or, *e.g.*, in the so-called resin tannins, on urea and formalin, on melamine and formalin or dicyanodiamide.

In the preparation of leather, animal skins, which contain a large amount of collagen, are treated with aqueous solutions of tannin. The phenolic hydroxyl groups of the tanning agents are thereby linked via hydrogen bonds to the collagen chains, leading to contraction and preservation of the skin materials. Finishing then follows, for instance, by dyeing the hides.

Chromatographic Conditions

TLC was carried out under standard conditions[15)], on 20 x 20 cm layers of silica gel GF_{254}. A two-stage technique has proved best for separating the phenols which come into consideration: The first stage for 10 cm at chamber saturation using toluene-ethylformate-formic acid (60 + 38 + 2) and, after 5 min intermediate drying, the 15 cm second stage using chloroform.

The phenols were localised by evaporating the solvent and marking the zones of reduced fluorescence in short-wave UV-light (254 nm). After then spraying with

Fast Blue Salt B (1% aqueous solution), the layer was dried, briefly exposed to ammonia and finally sprayed with conc. hydrochloric acid. Evaluation was performed a few minutes after reaction. The colour change on standing also furnished additional information. A 1% methanolic solution of the phenols 1–8 in Table 3 was used for comparison. Each time 1 μl was applied on the right side and 1 μl of the DESAGA colour test solution alongside.

Thermofractography

The thermal behaviour of free phenols and those linked to skin powder *via* hydrogen bonds is of interest for the interpretation of the thermofractograms of tannins and leather. The results of such experiments are summarised below in Table 3 under a) and b).

Table 3. Chromatographic and thermal behaviour of phenols a) free phenols on quartz wool, b) phenols partly linked to skin powder *via* hydrogen bonds

Substance	hRf-Range	Colour Reaction	Zones in TFG a) "free" in °C	b) "combined"
1) Phenol	50–60	red-violet	50–130	50–310 (max. 220–310)
2) Pyrocatechol	35–40	ochre yellow	50–170	50–310 (max. 230–310)
3) Resorcinol	30–35	red	50–190 (–250)[a]	70–330 (max. 250–300)[a]
4) Hydroquinone	27–32	yellow	50–200	80–310
5) Pyrogallol	20–30	orange	70–170 (200–260)[b] max. 240	100–200 (240–310)[b] max. 260
6) Methylphloro-glucinol	20–25	violet	–	–
7) Phloroglucinol	15–20	violet	130–230 (max. 190)	150–250 (max. 200)
8) Gallic acid	5–10	orange	170–240 (max. 230)	190–240[c] (max. 210)

[a] As a fragment from phloroglucinol.
[b] As a fragment from gallic acid.
[c] Very little undecomposed from skin powder.

The mono- and diphenols and pyrogallol appear undecomposed in various temperature ranges in the TFG. Phloroglucinol is partly decomposed into resorcinol. Gallic acid sublimes in the range 170–240 °C, undergoing partial decarboxylation into pyrogallol at over 200 °C.

Hydrogen bond formation with proteins (skin powder) displaces the zonal maxima clearly to higher temperature ranges. The behaviour of individual tannins and their structural units in the temperature gradient is given in Table 4:

Table 4. Thermolysis products and temperature of formation by TFG of uniforme tannins and their structure segments

Compound	Gallic acid [°C]			Pyrogallol [°C]			Phloroglucinol [°C]		
Gallic acid				200	240	260			
Butylgallate				150	220	250			
m-Digallic acid	160	265	280	180	270	280			
Trigalloylglucose	220	290	350	260	290	400			
Pentagalloylglucose	220	260	330	230	270	340			
Hamamelitannin	220	290	380	240	300	400			
Ellagic acid	220	330	400	250	330	370			
Gallocatechin				250	310	330	230	300	330

The temperature values mark begin, maximum and end of TFG-zones in °C

Catechin yielded the following thermolysis products (+ little, ++ much, +++ very much):

Phloroglucinol	280	320	380 ++
Methylphloroglucinol	280	310	360 +
Brenzcatechinol	270	310	380 +++
Resorcinol	300	310	360 +

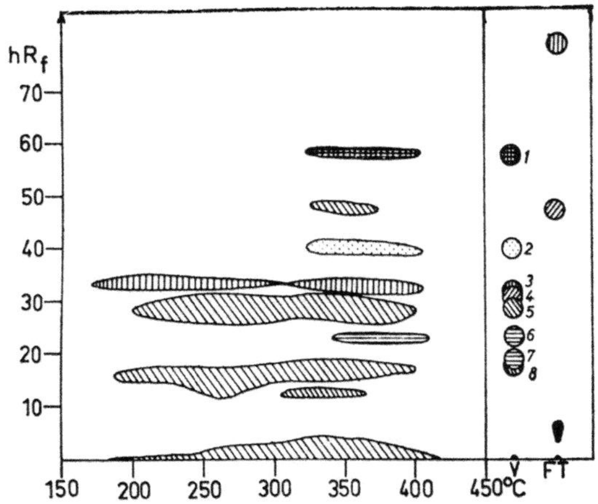

4

Fig. 8. Thermofractogram of 2 mg tannin with the typical gallotannin zones. The identification of the zones is possible with the reference solution V. 1) Phenol, 2) Pyrocatechol, 3) Resorcinol, 4) Hydroquinone, 5) Pyrogallol, 6) Methylphloroglucinol, 7) Phloroglucinol, 8) Gallic acid. FT Test mixture of three dyestuffs. Details see Table 3

As can be seen in Table 4, the esters and aromatic carboxylic acids generally suffer decomposition between 220 and 300 °C, yielding the corresponding phenols. The cleavage of C–C- and C–O–C-bonds takes place at above 300 °C. The influence of neighbouring groups is seen clearly here. In practice, it means that gallotannins and ellagitannins can be recognised in the TFG through the appearance of zones of gallic acid and pyrogallol. In contrast, phloroglucinol, resorcinol and pyrocatechol are the characteristic zones of the catechin tannins.

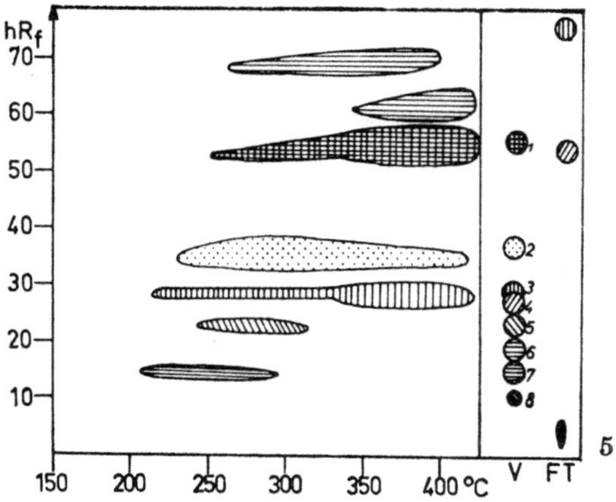

Fig. 9. Thermofractogram of 3,6 mg Catechu with the zones of a typical catechin tannin pyrolysate

Mixed tannins, such as gallocatechin, give the expected phenols. Investigations of medicinal plants containing tannins are best conducted with aqueous extracts. Fragments of water-insoluble polymers, *e.g.*, lignin and cellulose, do not then interfere.

The TFG of tannin, the aqueous extract of nut gall and a representative of the gallotannins, is depicted in Fig. 8. The TFG of the heart wood extract of Acacia catechu (drug: catechu), a catechin tannin, is in Fig. 9.

As already stated, the polyphenols are firmly bound to the collagen in leathers which have been subjected to vegetable tanning. In the TFG of a leather sample we may therefore expect that this bond will be broken and that fragmentation of both the tannin and the collagen will occur. Fig. 10 shows the TFG of an animal skin treated only with chromium salts (chrome-tanned leather). For comparison, Fig. 11 gives the TFG of a chrome leather that has been further tanned with mimosa extract; the characteristic, prominent zones of resorcinol (3) and pyrocatechol (2) show up clearly. The typical products of thermolysis of vegetable tanned and unfinished leathers are compiled in Table 5.

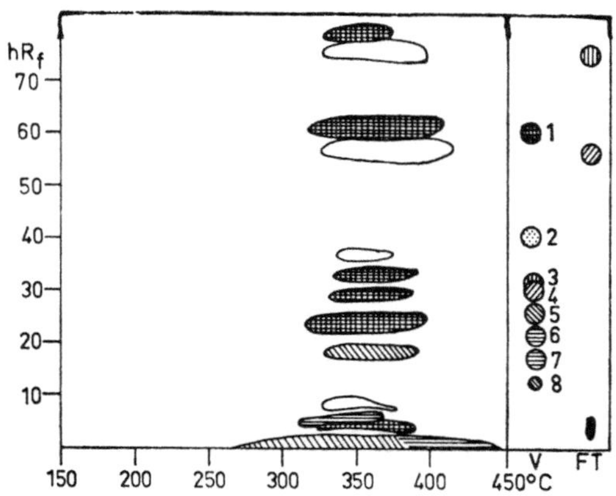

Fig. 10. Thermofractogram of 4 mg chrome-tanned leather. The zones are of collagen (animal skin). One cannot find typical phenol zones of a vegetable-tannin agent

Table 5. Characteristic thermolysis products from vegetable-tanned leather

Leather tanned with	Thermolysis products							Tannin class
	4-Vinyl-phenol	Pyro-catechol	Resor-cinol	Hydro-quinone	Pyro-gallol	Phloro-glucinol	Gallic acid	
Sumac(h) extract	+	+	+	–	+++	–	+	Gallotannins
Chestnut extract	+	+	+	–	+++	–	+	Gallotannins
Quebracho extract	+	+++	+++	–	++	+	–	Catechins
Mimosa extract	+	++	+++	–	++	+	–	
Chrome leather, further tanned with mimosa extract	+	++	+++	–	+	–	–	Inorganic + catechin

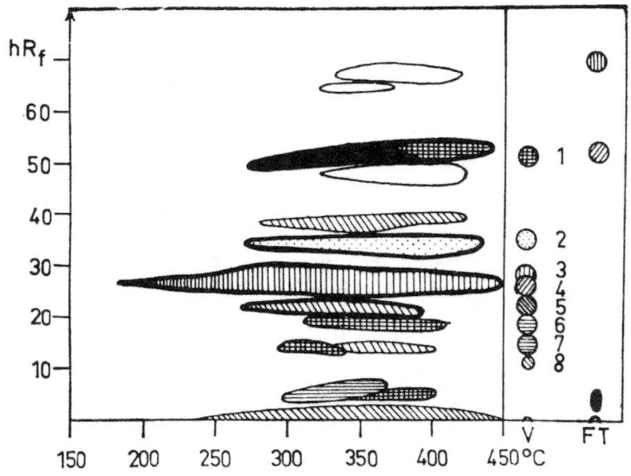

Fig. 11. Thermofractogram of vegetable-tanned (Mimosa extract leather). The heavy-circled zones are typical and are those of the tannin agent

Evaluation and Discussion

The thermofractogram of tannin (Fig. 8) shows a "double spindle" of gallic acid (8) in the hRf-region of 10–20. Its forward part (180–270 °C) comes from free or esterified gallic acid whereas the attached following zone is to be regarded as a thermolysis product of ellagic acid, especially as a strongly defined pyrogallol zone [(5), hRF of 40)] stands out in the temperature range 250–290 °C. Among free phenols, resorcinol [(3), 180–250 °C)] is present in the extract. Fragmentation of tannins possessing quite stable C–C- and C–O–C-bonds can be concluded from the appearance of the phenols, resorcinol (3), pyrocatechol (2), methylphloroglucinol (6) and phenol itself (1), as thermolysis products at above 300 °C with a maximum at about 350 °C.

In contrast to the TFG of tannin, the TFG of catechu (Fig. 9) shows no gallic acid and only very little pyrogallol. The well-defined zones of phloroglucinol (7), pyrocatechol (2) and phenol (1) (the first-named as a A-club, the last-named in the E-club form, and resorcinol (3) as a sort of double spindle). The violet stained zone above phenol (1) has not been identified. The furthest situated zone (hRf of 70) is guaiacol.

The phenolic fragmentation products of collagen have practically no influence on the evaluation of the TFG of leather. As Fig. 10 shows, they appear chiefly in the 325–400 °C range. This is clearly seen in the comparison with Fig. 11, of a leather sample subjected to further tanning with a mimosa extract. The powerful zones of resorcinol (3) and pyrocatechol (2) indicate vegetable tanning with a material containing a catechin tannin. With finished leathers, some of the materials used for treatment, *e.g.*, dyes, preserving agents and lipids, can likewise be recognised on the TFG. Little specialised experience is so far available here. This applies equally to the very promising domain of TFG of synthetic tanning agents. The small sample amount – 2–5 mg – makes the method just as interesting for a fast leather analysis, such as required in a criminological technical comparison, as for the study of consumer articles in the laboratory of a food chemist. Experiments so far have shown clear differences between the tannins of a class; the method ought therefore to be valuable also for future "fine structure analysis".

IV.3 Polysaccharides

Polymeric carbohydrates are substances of structural and storage importance for plants. Principally encountered here are cellulose, starch, hemicelluloses and, far less often, inulin. Celluloses in the form of fibre material for textiles (cotton) or as paper, and starch as food (flour) are indispensable for human beings.

The composition is usually deduced analytically by acid hydrolysis and identification of the resulting sugars. The thermal behaviour of the polysaccharides is important for solving a number of problems. Newer investigations in this field have been based mostly on pyrolysis gas chromatography. In the customary temperature range of TFG, carbohydrates yield only about 5% volatile products and 80% condensable on the thin layer, however[128]. Interesting in the sphere of foodstuffs and semiluxuries are the "non-enzymatic browning reactions" of carbohydrates (Maillard reaction) which yield typical flavours.

Thermofractography

The polysaccharides investigated begin to undergo thermolysis between 220 and 250 °C. The principal cellulose and starch zones appear between 270 and 370 °C. A series of E-club shaped zones (about 10) is obtained which can be identified through their fluorescence quenching in UV_{254} and various colour reactions.

TFG-Conditions:

Sample size: about 3–4 mg
Temperature range: from 200 to 450 °C at 8 °C per minute
Displacement: 6 mm per minute, carrier gas: 15 ml per minute

Thin layer: silica gel GF_{254}
Developing solvent: chloroform-ethanol (93 + 7)
Run: 15 cm at chamber saturation
Detection:
1. UV_{254} (zones marked).
2. Fast Blue Salt B reagent; then sprayed with 36% hydrochloric acid, observing colour changes.
3. After 2, sprayed with saturated o-dianisidine in glacial acid, observing colour changes of the aldehydes.

Thermofractogram of Cellulose

Differences between the thermofractograms of the various polysaccharides were ascertained in exploratory experiments, although admittedly not in the chief zones. Only cellulose is considered here as a representative of the polysaccharides. The two cyclopentadiones (4, 5) with their characteristic reactions with Fast Blue Salt B (see Table 6) can be regarded as "guide" substances in the thermofractogram (Fig. 12). Zones of furfuraldehyde (1) and its derivatives (2, 7) may also be mentioned. The other polysaccharides and saccharides show these zones also in varying intensities. Cotton can be identified in blended fabric through these TFG-zones (Fig. 12).

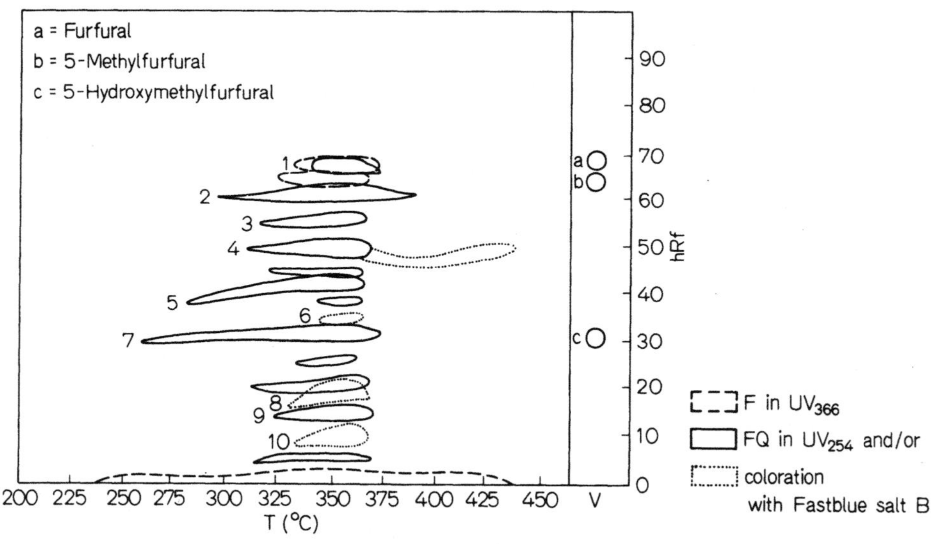

Fig. 12. Thermofractogram of 4 mg cellulose. Numbers of zones and colours see Table 6. V = reference solution: a) furfural, b) 5-methylfurfural, c) 5-hydroxymethylfurfural

IV.4 Proteins, Blended Fabrics

The natural protein fibres, such as sheep's wool and silk, contrast with the fibres prepared from animal or plant proteins. Exploratory tests have shown that the thermolysis of these natural fibres begins at about 250° and continues almost to 400 °C.

Table 6. The 10 TFG zones of cellulose and their detection with different reagents

No	hRf	Substance	UV_{254}	Fast blue salt B-reagent[158]	Anisaldehyde-sulfuric acid-reag.[15]	o-Dianisidine-reg.[15]
1	85	Furfural	+	–	Blue	Faint yellow, faint violet
2	80	Methylfurfural	+	Yellow yellow-orange	–	Yellow
3	60	–	+	Orange red	Yellow	Faint green
4	55	3-Methylcyclopentane-1,3-dione	+	Brownish red	Blue-green	–
5	50	Cyclopentane-1,2-dione	+	Burgundy red blue	Moss green	–
6	40	–	–	–	Yellow olive	Red orange
7	35	5-Hydroxymethylfurfural	+	Faint yellow	Blue green	Orange yellow
8	20	–	–	Red orange	–	–
9	15	–	+	Violet	Olive green	Faint green
10	10	–	–	Orange reddish brown	–	–

Table 7. Typical zones in the TFG of fibre materials

hRf-values	UV_{254}	Fast blue salt B reaction[158]	Material	Observations
60–63	+	Violet-blue (4-vinylphenol)	Silk	Marked zone
57–63	+	Yellow-orange, red-violet (phenol)	Silk, wool	Marked zone
55–60	+	Burgundy-red-blue (cyclopentane-1,2-dione)	Cellulose	Marked zone
45–50	+	Faint yellow (hydroxymethylfurfural)	Cellulose	
48–50	+	Purple	Silk, wool	
38–42	+	(Yellow in daylight)	Silk	Marked zone
34–37	+	Red-violet, orange-brown	Silk	
32–36	–	Blue purple	PAN	
25–28	–	Orange, brown-red	Cellulose	
22–27	+	Red-violet, orange-brown	Silk, wool	Marked zone
15–18	–	Red-violet	Wool	
13–15	+	Orange, red-brown	Polyester	
5–45	+	6 Zones, faint yellow	Polyester	Marked zones
7–10	–	Red-violet, brown (hexylamine)	Polyamide	

(+) Means: fluorescence quenching when inspected in UV_{254} light.
(–)Means: absence of this property of fluorescence quenching.

Sample sizes for TFG are of the order of 15 mg. The zones are identified through fluorescence in UV_{366}, fluorescence quenching in UV_{254} and the reaction with ninhydrin. Separation is carried out on silica gel GF_{254} using chloroform-methanol-conc.ammonia (80 + 20 + 2). The sometimes quite varied zones from the different protein fibres cannot yet be identified. Nevertheless, the typical zones given in Table 7 permit the components of blended fabrics to be identified.

V Thermofractography of Synthetic Polymers

V.1 Phenol Resins[129, 130]

V.1.1 Preliminary Remarks

Phenol resins are polycondensation products of phenols with carbonyl compounds, mostly formaldehyde[5, 18, 131–133]. Condensation under acid conditions with a deficiency of formaldehyde yields soluble, thermoplastic phenol resins. Essentially the benzene nuclei are linked by methylene bridges in o- and p-position. Some of these positions remain admittedly unsubstituted. The condensation polymers thus obtained are called *Novolaks.* In contrast, polycondensation in basic medium and with excess of formaldehyde regularly gives soluble, thermoplastic hydroxymethylphenols, with one or more nuclei. These contain reactive nuclear hydrogen atoms as well as the methylol groups which can undergo further condensation. These condensation products, termed Phenol *Resols,* are thus able to react further on heating, forming cross-links. This further condensation, "self-hardening", is a contrast to the novolaks. After passing through the *Resitol*-stage, they yield close-meshed cross-linked three-dimensional polymers which are completely insoluble and infusible called *Resites.*

V.1.2 State of the Analysis of Phenol Resins

Phenol resins are generally identified by liberating phenol from the thermosetting sample by heat or hydrolysis and then applying rapid tests for it[5,6,24,91]. However, detailed information about the presumed type of phenoplast can be obtained only by its thermal breakdown into the structural units and their separation and identification by GC[49,52,58,91,93] or TLC[91,93]. This led to the discovery that the pyrolysis products of higher alkylphenol-formaldehyde condensation polymers do not give conclusive chromatograms[91]. The technique described above moreover fails to distinguish between novolak and resol types.

V.1.3 TFG-Analysis of Phenoplasts[129,130]

In the meantime, thermofractography has provided a convincing solution of the problems mentioned above. Our experience leads us to recommend the procedure described below.

V.1.3.1 Chromatographic Conditions

Instructions for the TLC of the phenols expected as products of thermolysis were abundantly available already[15]. Preliminary experiments showed that the phenol mixtures to be expected were best separated by double development using benzene in a run of 15 cm. For speed and reproducibility this solvent gives the best results although superposition of phenol with m- and p-cresols and also some higher alkylphenols (cf. Table 8) is unavoidable. Partition-TLC on layers impregnated with formamide has been used successfully to separate these critical homologous or isomeric phenols[91,129,130]. We have studied thoroughly the scope of application and performance of this chromatographic system, including also the impregnation technique[113]. However, formamide-impregnated layers are not very suitable for collecting phenolic products from thermolysis of phenol resins (nor, as will be shown, from epoxide resins); the hot vapour stream of carrier gas and thermolysis products evaporates the formamide in the vicinity of the start band. Further, formamide layers have poor mechanical strength since the adhesion to the supporting plate is marked-

Table 8. TLC-conditions and detection reactions for degradation products of phenol resins

Thermolysis product	Phenoplast type	Solvent, layers, hRf-values I	II	Detection[158]
Phenol	A, B, D, E	19–20 Yellow	18–20 Red	Sprayed with
o-Cresol	A, B, D, E	48–50 Green	28–30 Violet	5–10%
m-Cresol	D	38–40 Green	18–20 Violet	aqueous
p-Cresol	A	38–40 Green	18–20 Brown	Fast blue
2.3-Xylenol	D	66–68 Yellow	28–30 Blue	Salt B.
2.4-Xylenol	B	74–76 Red	28–30 Brown	Layer allowed
2.5-Xylenol	D	73–75 Orange	38–45 Violet	to dry. Expos-
2.6-Xylenol	A, B, E	88–90 Green	41–43 Violet	ed to ammonia
3.4-Xylenol	D	54–56 Red	17–17 Violet*	vapour.
3.5-Xylenol	D	56–58 Green	18–20 Violet	Sprayed with
2.4.6-Trimethylphenol	B	90–94 Faint violet	43–45 Reddish	conc. hydro-
2.4.5-Trimethylphenol		85–87 Brown	28–32 Brown	chloric acid;
2.3.4.6-Tetramethylphenol	B	–	74–78 Violet	colour change
4-tert. Butylphenol	C	88–90 Yellow	23–25 Brownviolet	observed.
2-Methyl-4-tert. butylphenol	C	Front yellow	35–37 Brown	
2.6-Dimethyl-4-tert-butylphenol	C	–	48–52 Brown	Fluorescence
2.3.6-Trimethyl-4-tert. butylphenol	C	–	68–72 Brown	quenching in
Diphenylmethanes	A, B, C, D	Front	Front	UV_{254}

A phenol-novolak.
B o-cresol-novolak.
C 4-tert-butyl-phenol-novolak.
D xylenol-novolak.
E phenol-resol.

I Silica gel HF_{254}, 40% impregnated with formamide; dichloromethane-cyclohexane (55 + 45); chamber saturation, one 15 cm run.
II Silica gel HF_{254}, benzene; chamber saturation; two 15 cm runs.

ly reduced. The TFG-zones obtained after chromatography using benzene normally suffice to differentiate the various novolak and resol types. The TFG-zones in question are located first by ringing the areas of reduced fluorescence on inspection in UV_{254}. Fast Blue Salt B as a 5–10% aqueous solution has proved suitable for visualising the phenol zones. The sprayed reagent couples with the phenols on the fractogram to give variously coloured azo dyes. The coupling reaction is enhanced by exposing the layer to ammonia vapour. Any reddening of the background which this brings about can be eliminated by then spraying with conc. hydrochloric acid. The colour change of certain TFG-zones which then occurs is noteworthy and typical for individual phenols. The hydrochloric acid sprayed on also helps preserve the colours of the zones.

Table 8 contains a survey of the chromatographic conditions and colour reactions of the phenols encountered as thermolysis products from phenol resins.

V.1.3.2 Examples of Analyses and Results from TFG

Among the many industrially prepared phenol resins, novolak types from phenol, o-cresol, 4-tert-butylphenol and a technical xylenol fraction were studied. The comparison with a phenol-resol also appeared attractive. The thermofractograms obtained showed characteristic differences and also certain features in common, which are in general typical for phenoplasts. The thermal fission of the samples analysed here began always at about 150 °C and was usually not complete even at 450 °C, the maximum temperature which can be attained by the TFG apparatus. This conclusion is based chiefly on the relevant TFG-zones at 450 °C which, without exception, have the E-club form. It can also been seen from the thermofractograms that the thermolysis products of all phenoplasts always contain the regenerated corresponding starting compounds along with higher homologous and isomeric phenols. This indicates that bond fission takes place preferentially at the methylene bridges linking the phenyl groups. Alkyl substituents in the benzene nuclei evidently require a higher fission energy since they are not detached under the given thermal conditions. Secondary thermolysis products are not encountered on account of the rapid removal from the thermolysis region and the inert gas atmosphere. All phenoplasts, especially those from condensation polymers prepared from complex technical starting mixtures, are distinguished by their more or less large content of unreacted monomers. This is easily demonstrated through TFG-analysis since the free monomers in the phenoplast are transferred to the thin layer before the thermal energy for depolymerisation into the same starting components has been reached. This is shown on the thermofractogram through a long drawn out TFG-zone with the form of a linked (cf. Fig. 13) or separated (cf. Fig. 14) double spindle in the same hRF-region. Mixtures of diphenylmethanes occur in the vicinity of the front beyond ca. 400 °C in all thermofractograms. These are manifestly formed by homolytic fission of phenyl groups joined by an ether linkage.

The identification and differentiation of two types of phenol resin, nevertheless closely related, are illustrated below with the example of TFG-analyses of phenol-novolak and phenol-resol.

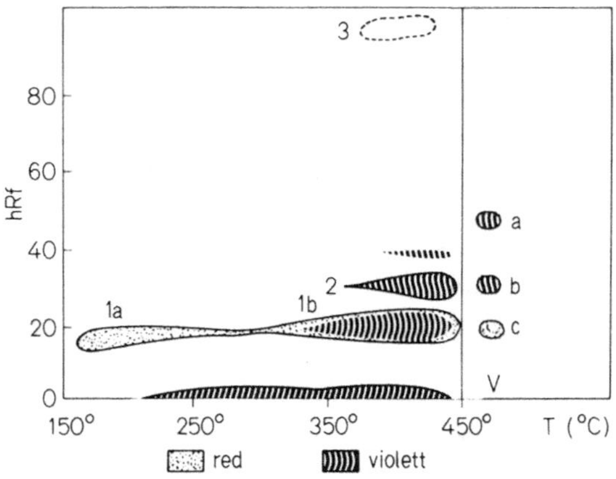

Fig. 13. Experimental details see Table 8. Reference mixture V: a) 2.6-xylenol, b) o-cresol c) phenol

V.1.3.2.1 TFG of Phenol-Formaldehyde-Novolak

The TFG of phenol-novolak (see Fig. 13) shows in the range 150–300 °C the free starting phenol which occurs in the A-club form as a result of its appreciable vapour pressure at room temperature. To this is attached from 300 °C a phenol zone as E-club (1 b), derived from the thermal degradation of the condensation polymer into its starting component. The TFG-zones of the phenol yield together the double spindle-form which is characteristic for the novolak types. p-Cresol can be detected in the thermolytically yielded later spindle section (1 b), after TLC-TLC-liquid transfer to a silica gel layer impregnated with formamide and then GC. o-Cresol (2) appears in the E-club form above the phenol zone (1) but no 2.6-xylenol can be detected although it is a theoretically possible thermolysis product. In the front region between 380 and 430 °C a zone of fluorescence quenching occurs (3) due to a mixture of successively methylated diphenylmethanes. These are formed in the thermal fragmentation only of novolaks. They are not yielded by resols, nor are high molecular oligomer condensation products found in the start band.

V.1.3.2.2 TFG of Phenol-Resol

The thermofractogram of phenol-resol shows distinct variations in the number and form of the zones: this contrasts with the TFG of the novolaks. Phenol (1) appears as a prominent A-club in the region of 150–250 °C, tapering to 450 °C. That it begins as an A-club zone in the lower temperature range shows that it comes from free phenol which did not take part in the polycondensation. This assumption is supported by the fact that the first half-spindle of o-cresol (2 a) appears above the phenol zone in the same temperature region. From the surface areas of the zones it can be estimated roughly that the starting mixture consisted mainly of phenol with a little o-cresol. The relatively large amount of unreacted phenol justifies the conclusion that the resol has a low degree of polymerisation. This also explains why the TFG-zones of free and combined phenol merge into each other. Another striking feature is the absence in the start region from 250 °C of condensation products of high molecular weight. This indicates that the products formed through further con-

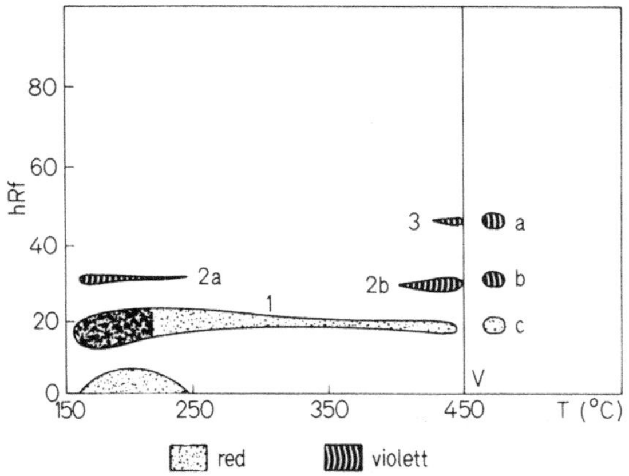

Fig. 14. Experimental details see Table 8. Reference mixture V: see Fig. 13

densation *via* methylene or ether bridges have completely decomposed into the starting materials. The weak zones of o-cresol (2b) and 2,6-xylenol (3) encourage the assumption that they arise only from phenols linked *via* methylene bridges.

V.1.3.3 Discussion and Evaluation

Thermofractograms of novolaks o-cresol, 4-tert-butylphenol and a technical xylenol were recorded as well as those novolak and resol types described above in detail[129, 130] Typical TFG zones were found in each case (cf. Table 8). The formation of 2,3,4,6-tetramethylphenol above 300 °C together with the expected fission products o-cresol, 2,3-xylenol, 2,6-xylenol and mesitol, is a conspicuous feature of the o-cresol-novolak. Its presence can be explained only when one assumes that condensation also in the m-position as well as the customary formation of methylene bridges in o- and p-positions takes place in the presence of acid catalysts. Incidentally, this TFG-zone is characteristic for the o-cresol type. The TFG of the phenolic resin based on a technical xylenol mixture is also characteristic for novolaks. The front part of the A-club, beginning at about 150 °C, shows clearly that the starting mixture consisted of phenol, much m-cresol and 3,5-xylenol and little 2,3-xylenol. The thermal breakdown does not begin until ca. 350 °C and also liberates various methylated phenols. The absence of a 2,6-xylenol zone distinguishes this resin type. Altogether, the number, shape and arrangement of the TFG zones obtained here can serve to yield "finger prints" for characterising the sample. The TFG of a "pure" 4-tert-butyl-novolak also furnishes clear features for classification. It is distinguished especially through a number of phenol zones stained brown with Fast Blue Salt B. These were identified in the order of their appearance (with increasing hRf-value) as the starting component 4-tert-butylphenol, accompanied by some higher methylated homologues up to trimethyl-4-tert-butylphenol. Here, too, this can be explained only on the assumption of linkage in the m-position. A technical 4-tert-butylphenol resin differed from the pure product in showing a large number of phenol zones in the lower temperature region of 150–300 °C. Comparison of the thermofractograms shows impressively the differences in the composition of the starting components. This is at the same time evidence of the possibility of distinguishing between starting phenols and thermolytically yielded fission products through TFG-analysis. This contributes decisively to a detailed characterisation of the various phenoplast types. A simple analysis scheme for phenol resins based on thermofractography is given on page 37.

V.2 Epoxy Resins[136, 137]

V.2.1 Introduction[131, 132]

Epoxy resins are the products of the basically catalysed reaction between polyfunctional hydroxy compounds and epichlorohydrin. Starting materials include polyfunctional phenols and phenoplasts, and aliphatic and alicyclic polyols and polycarboxylic acids. The first products of their polycondensation with epichlorohydrin are liquid or solid thermoplastic resins with variable contents of epoxide end groups, capable of further reaction. These are the so-called unhardened epoxy resins or primary epoxy resin products. The technically most important representatives are the polycondensation products from bisphenol A and epichlorophydrin. Depending on the degree of condensation, they have molecular weights from 450 to 4000, corresponding to an oligomer number from 1 to 12. The primary products from phenol novolaks with more than 2 reactive epoxide end groups per molecule are highly important since they yield hardened products resistant to high temperature. The content of reactive epoxide groups confers on the primary epoxy resin products a special tendency to polyaddition with cross-linking agents or hardeners. These are generally compounds of low molecular weight containing active hydrogen atoms, which open the oxirane rings of the bi- or oligofunctional glycidyl ethers or esters in a nucleophilic substitution leading to formation of secondary alcohol groups. Aromatic diamines, aliphatic polyamines and the anhydrides of polycarboxylic acids yield hardened systems with favourable material properties. This polyaddition reaction between primary products and hardeners gives incidentally no volatile subsidiary products and the extensive cross-linking gives insoluble thermosetting materials resistant to mechanical and chemical attack and with good electrical properties.

V.2.2 Methods of Identification

Like phenol resins, epoxy resins can be identified rapidly with the help of a number of relatively simple, yet reliable chemical spot tests[92, 138], but they serve mostly only for establishing the class of polymer. More subtle analytical techniques are required to characterise technical epoxy resin formulation, especially in the presence of additives, fillers or perhaps even of spurious materials. The decisive breakthrough in the systematic characterisation of epoxy resins came with the introduction of thermal degradation procedures into polymer analysis. The polymer composition can be established beyond doubt from the separated and identified products of thermal degradation. The problems of separation and detection of these products have been solved in practice by GLC[72, 74, 139–142], TLC[86–88, 92, 93, 143], MS[76, 77, 143–147] and

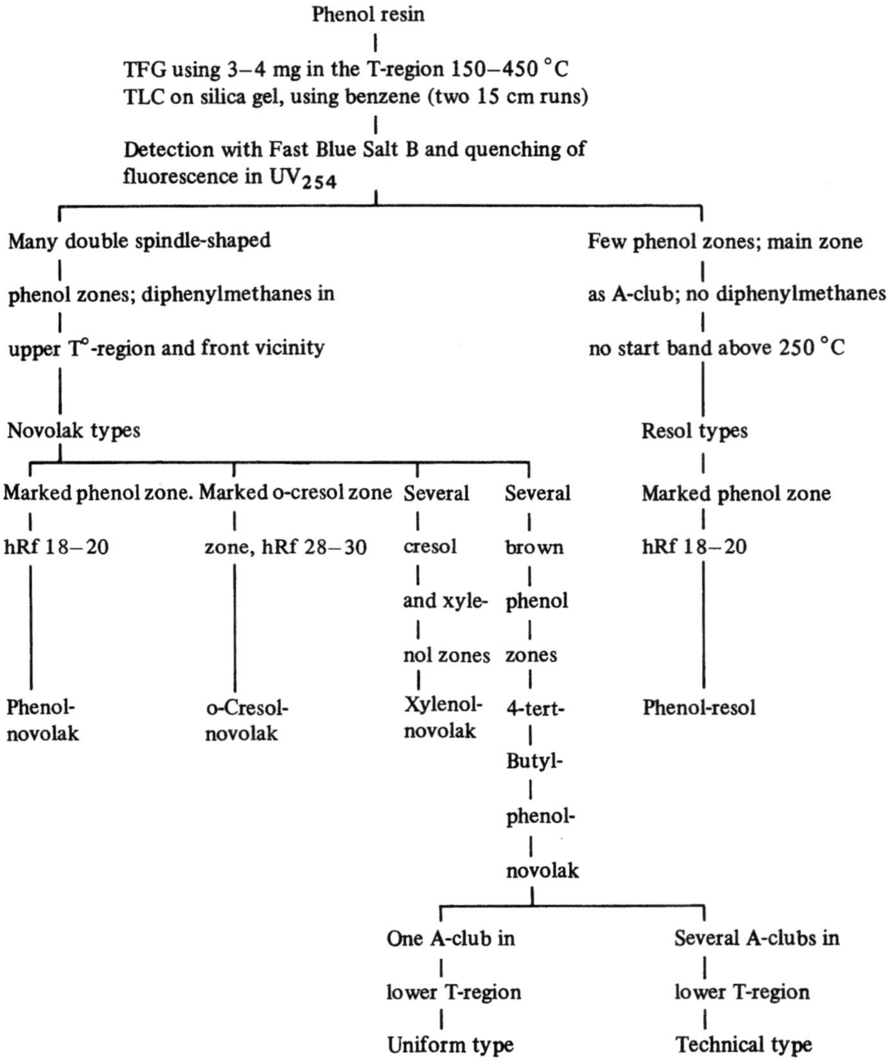

Scheme 2. Identification key for TFG-analysis of phenol resins

IR-spectroscopy[40,41,43,44,148]. The coupled procedure pyrolysis-GC-MS[83,149–151] has proved to be an especially rich source of information. These methods retain, nevertheless, certain shortcomings. The restriction to isothermal degradation and the high pyrolysis temperatures cause lower epoxy resin oligomers to be prematurely decomposed and hence escape detection. Consequently, no statement can be made about the degree of condensation of the primary products. In addition, free and re-formed monomers cannot be distinguished. The hardening components of thermostable epoxy resins can be identified only occasionally. There is much ground to be made good in analytical methods just in this field.

Table 9. TLC of thermolysis products of epoxy resins. Standard conditions[15], Chamber saturation, basic components, layers, solvents, hRf-regions and detections

Thermolysis products	Basic components	hRf-Values I	II	Colour	III	Colour	IV	V	Detection
Phenol	A B C	35–40	38–43	Carmine	35–40	Orange			Fast Blue Salt B - Reagent 91[15], modified[158]
o-Cresol	A B C	48–52	45–50	Violet red	57–62	Green			
p-Cresol	A B C	35–40	38–43	Brown	50–53	Yellow orange			
4-Ethylphenol	A	35–40	38–43	Brown	70–72	Light red			
2-Ethylphenol	A	58–62	54–58	Blue violet	77–82	Red			
4-Isopropylphenol	A	38–42	48–51	Light brown	73–78	Yellow			
4-Isopropenylphenol	A	38–42	48–51	Dark blue	63–67	Blue			
Bisphenol A	A	8–12	17–20	Dark brown	10–12	Yellow brown			
Bis2HPM	B C	28–30	20–25	Blue gray	35–40	Gray			
Bis4HPM	B C	15–18	10–15	Red brown	8–10	Brown			
2.4-Dimethylphenol	A B C	50–55	45–50	Red	65–70	Red brown			
2.6-Dimethylphenol	B C	68–72	60–63	Violet	80–83	Green			
Mesitol	B C	68–72	60–63	Slight pink	90–93	Slight violet			
Bis4HPM-DGE	B	50–60	45–50		75–85				FQ in UV_{254}
Bisphenol A-DGE	A	50–60	50–60		90–95				
Benzofuranes	A B C	78–80	70–75		90–100				FQ in UV_{254} and F in UV_{366}
Benzo-γ-pyranes	A B C	93–98	80–85		90–100				
Xanthenes	B C	85–90	85–90		90–100				
DDM	a	18–20		Yellow brown					F in UV_{366} with Fluram[157]
Aniline	a	45–48		Yellow green					
N-Methylaniline	a	64–66		Light green					
N-Methyltoluidine	a	64–66		Light green					
Tetramethyl-DDM	a	45–48		Red violet					K_2PtJ_6 R.141[15]
HPA	b D	16–20		Slight violet		Detection	83–88	Yellow	Bromocresolgreen

HPS	b D	3–6	Slight violet	Fast Blue Salt B	36–40	Yellow	R 31[15]
Maleic anhydride	b D				17–20	Yellow	
Cyclohexanone	b D	62–65	Slight blue				Fast Blue Salt B[158]
			Violet red				Anisaldehyde-H_2SO_4
Glycerin	b D					28–30	R 15[15]
						Violet red	

Basic components: A bisphenol, B phenol-novolac, C o-cresol-novolac, D hexahydrophthalicdiglycidylester.
Hardeners: a 4.4'-diaminodiphenylmethane (DDM), b hexahydrophthalic anhydride (HPA).
Layers and solvents: I Silica gel 60 HF_{254}-GF_{254} (1 + 1) with chloroform 2 x 10 cm with 5 min drying in between, CS II Silica gel 60 HF_{254} with dichloromethane-Ether (95 + 5) 1 x 10 cm, CS.
III Silica gel 60 HF_{254} 90% formamide impregnated with toluene-dichloromethane (50 + 50) 1 x 10 cm, CS IV Kieselguhr G 25% polyethyleneglycole 1000-impregnated, with upper phase diisopropylether-water – formic acid-polyethyleneglycole 1000 (90 + 3 + 7 + 2) 1 x 10 cm, CS V Silica gel 60 HF_{254}-GF_{254} (1 + 1) with n-butanol, ammonia saturated, 1 x 10 cm, CS.

V.2.3 TFG-Analysis of the Basic Components of Epoxy Resins[136)]

V.2.3.1 TLC-Conditions

The chromatographic conditions had to aim at the best separation of the alkylphenol zones characterising the individual epoxy resin types. However, the solvent benzene, used in the TLC of thermolysis products of phenoplasts, was unsuitable here. To characterise technically important polyepoxides, the detection additionally of some diphenol TFG-zones is necessary (cf. Figs. 15–16). Double development using chloroform in two runs of 15 cm separates satisfactorily 2- and 4-alkylphenols and all important diphenols. Only where assignment of the zones was uncertain or several phenols overlapped in a TFG-zone, was it necessary, as previously described[130, 113)] to fall back on partition chromatography on formamide-impregnated silica gel layers after microtransfer. After removing solvent traces from the layer in a stream of hot air, zones fluorescing in UV_{366} and zones of quenched fluorescence in UV_{254} were marked. As is usual with phenols, the zones were stained through azo coupling with the Fast Blue Salt B reagent, mentioned above. After carrying out the procedure described in Section V.1.3.1., characteristic colours were obtained for the individual TFG-zones of the phenolic degradation products. In most cases, these permit the particular phenols to be identified. Together with co-chromatography of suitable reference mixtures and the typical shapes and patterns of the TFG-zones, the identification of the basic components is adequately ensured. The experimental details and relevant TLC-data are to be found in Table 9.

V.2.3.2 Discussion of the Results

Our analytical results are based on the study of some unhardened epoxy resins derived from phenol-novolak, o-cresol-novolak and bisphenol A. Further study was devoted to the TFG-behaviour of hardened bisphenol A-epoxy resins, "cured" with 4,4'-diaminodiphenylmethane (DDM), triethylenetetramine (TETA) and hexahydrophthalic anhydride (HPA) as cross-linking agents, and to a cycloaliphatic system hardened with HPA. The following observations were made: Unhardened bisphenol A- and phenol-novolak-epoxy resins, and HPA-hardened bisphenol A- and cycloaliphatic epoxy resins begin to decompose at 200 °C (cf. Fig. 15). In contrast, thermal breakdown of unhardened o-cresol-novolak-, DDM- and TETA-hardened bisphenol A-epoxy resins does not start until above 250 °C (cf. Fig. 16). The type and number of thermolysis products of the individual types differ considerably also. Unhardened phenol-novolak- and bisphenol A-systems are easily recognisable through the transferred oligomeric diglycidyl ethers in the lower temperature region (zone 6 in Fig. 15). Specific alkylphenol zones then permit further differentiation. HPA-hardened bisphenol A- and cycloaliphatic polyepoxides are easily distinguishable through the detection of phenol zones with the former. The unhardened polyadduct on the basis of o-cresol is the only epoxy resin which does not decompose until above 300 °C. There is no difficulty in identifying the DDM-hardened bisphenol A-types because the TETA-hardened type yields no amine-containing zones in the TFG. Two thermofractograms of bisphenol A-systems are reproduced below to

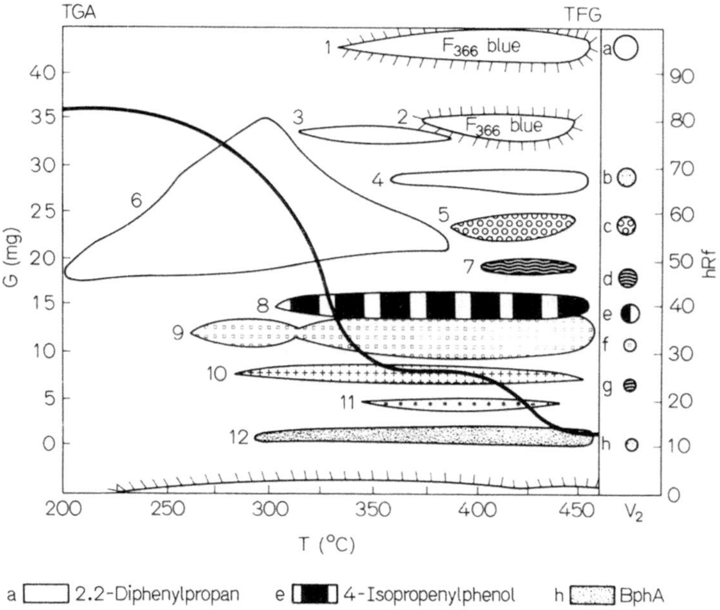

Fig. 15. TFG of 2 mg unhardened, low-condensated bisphenol A-epoxy resin. Experimental details see Table 9. Reference mixture V_2: a) 2.2-diphenylpropane, b) 2.6-xylenol, c) 2-ethylphenol, d) o-cresol, e) 4-isopropenylphenol, f) phenol, g) bis[2-hydroxyphenyl]methane, h) bisphenol A

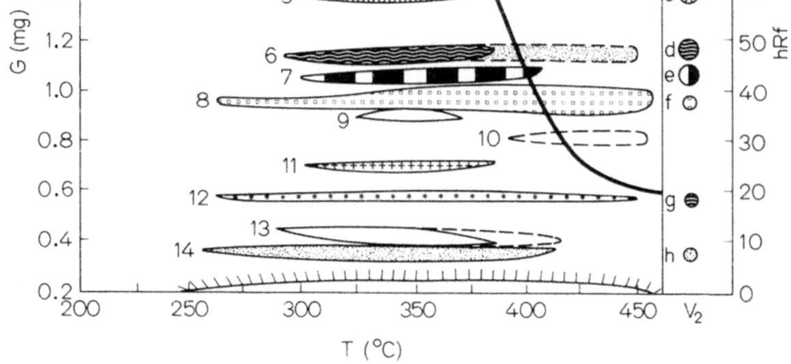

Fig. 16. TFG of DDM-hardened bisphenol A-epoxy resin. Experimental details see table 9. Reference mixture V_2 see Fig. 15

Epoxy resin

TFG with 2–3 mg in temperature region 200–450 °C
TLC with solvent chloroform 2 x 10 cm with 5 min drying in between
detection of TFG-zones with Fast Blue Salt B (see Table 8, also)

- TFG-zones from 200 °C
  - Unhardened bisphenol-A- and phenol-novolak-, hardened BA- and cycloaliphatic EP-resins (HPA)
    - Detection of DGE in hRf-region 50–60 by fluorescence-quenching in UV_{254}, from 200 °C
      - Positive
        - Comparison of hRf-values and colours of TFG-zones with authentic phenols
          - hRf
            8–13 bisphenol A
            30–40 phenol
            40–45 4-isopropenylphenol
            45–50 o-cresol
            55–60 2-ethylphenol
            - Unhardened BA-EP-resins
              - BA-DGE-zone
                - hRf 50–60
                  - Strong / Slight
                    Lowly / Highly
                    Condensed resin type
          - hRf
            15–20 Bis4HPM
            25–30 Bis2HPM
            38–40 phenol
            50–55 o-cresol
            67–70 2.6-DMP
            - Unhardened phe-nol-novolak-EP resins
              - Bis4HPM-DGE
                - zone
                  - hRf 50–60, from 200 °C
                    - Strong / Slight
                      Lowly / Highly
                      Condensed resin
      - Negative
        - Detection of faint violet acid- and anhydride-zones near the start from 200 °C
          - HPA-hardened EP-resins detection of phenols
            - Negative
              - Hardened cylco aliphatic EP-re-sins
            - Positive
              hRf
              8–13 bisphenol A
              30–40 phenol
              40–45 4-isopro-penylph.
              45–50 o-cresol
              55–60 4-ethylph.
              - Hardened BA-EP-resins

Scheme 3. Identification key to TFG-analysis of the basic components of epoxy resins

TFG-zones from 250 °C

Unhardened o-cresol-novolak-, hardened BA-EP-resins, hardener DDM and TETA

Start of thermolysis above 300 °C

Positive

Comparison of hRf-values and colours of TFG-zones with authentic phenols

hRf

15–20 Bis4HPM
25–30 Bis2HPM
38–40 phenol (slight)
50–55 o-cresol (strong)
67–70 2.6-DMP

Unhardened o-cresol-novolak-EP-resins

Negative

Detection of amine-zones in the middle and lower hRf-region with fluorescamine

- Positive
  hRf

  8–13 Bisphenol A
  10–15 DDM
  30–40 phenol
  40–45 4-isopropenylph.
  45–50 o-cresol
  45–50 aniline
  55–60 2-ethylphenol

  DDM-hardened BA-EP-resins

- Negative

  TETA-hardened BA-EP-resins

illustrate the differences in TFG-behaviour between unhardened and hardened epoxy resins.

The relevant TGA-curves are placed under the two thermofractograms to give a quantitative idea of the extent of thermal breakdown. Further, Scheme 3 contains the TFG-results, described above, schematically presented as an analytical key.

V.2.4 TFG-Analysis of Hardening Agents of Epoxy Resins

Special procedures for identifying and distinguishing the hardeners of epoxy resins are hardly known at the moment. They are generally identified within the analytical scheme for the basic components[41,86–88,92,140–145]. Although these methods have yielded indisputable successes they have grave shortcomings. Often only the class of hardener can be recognised but without differentiating the hardeners among themselves[41,86–88,92,140,149]. Further, there is so far no known method for detecting the polyethylenepolyamine hardeners[41,87,140,151]. It was therefore of supreme interest to supplement and extend the techniques aiming at specific analysis of hardeners.

V.2.4.1 TLC-Conditions

Some tested procedures exist for the TLC of the compound classes coming into consideration as hardeners. We have used the ternary mixture toluene-acetone-25%ammonia (80 + 20 + 1), a modification of a common solvent system[152], in a single 10 cm run for aromatic diamines. Solvent systems without ammonia can also be used (cf. Table 12). More drastic conditions are necessary to separate the polyethylenepolyamines, however. The often quoted, highly polar solvent ethanol-25%ammonia (33 + 66)[153,154] in a single 10 cm run is very good (cf. Table 11). Excellent TLC-conditions are known for the separation of homologous and isomeric aliphatic and aromatic dicarboxylic acids[155,156]. Satisfactory separation can often be accomplished by adsorption chromatography on silica gel layers using solvents containing acid or ammonia[130,155]. We have thus used the system chloroform-ethyl acetate-formic acid (49 + 49 + 2) in successful TLC-group analysis of the anhydride hardeners[130,137]. Difficult separations of homologous or isomeric dicarboxylic acids are better achieved through partition chromatography. For this, a single 15 cm run is used on kieselguhr layers impregnated with 25% polyethylene glycol 1000, and as solvent system, the upperphase of diisopropyl ether-formic acid-water-polyethylene glycol 1000 (90 + 7 + 3 +2) (cf. Table 13)[156]. The hardener fragments from thermolysis of epoxy resins hardened with dicyanodiamide, are separated very well using pyridine-benzene-acetonitrile-water (25 + 40 + 30 + 3), normal saturation and a run of 10 cm[102,137] (cf. Table 14). Several reagents for specific groups are suitable for detecting the various groups of hardeners and their fragments. They are used in accordance with the information in Table 10. This table contains also experimental conditions for identification of the individual hardeners through alkali fission.

Table 10. TAS- and TLC-conditions for alkaline fusion (*A*) and acidic desintegration (*SA*) of hardened epoxy resins. (Length of application 20 mm, carrier gas: 10 ml N_2/min)

Hardener	TAS-Conditions				TLC-Conditions	
	Amount (mg)	Technique	Temperature [°C]	Time of sojourn in sec.	Layers and solvents	Detections
Aromatic diamines	1–2	A	250	60	Silica gel HF_{254}-GF_{254}(1 + 1), toluene-acetone-ammonia 25% (80 + 20 + 1), 1 x 10 cm, CS	Fluorescamine 0.02% in Acetone[157]
Aliphatic polyamines	5–8	A	300	90	Silica gel HF_{254}-GF_{254} (1 + 1) ethanol-ammonia 25% (33 + 66), 1 x 10 cm, CS	Ninhydrine[15] [Reag. 175] 1% in n-butanol, fluorescamine[157]
Dicyanodiamide	5	A	250	60	Silica gel HF_{254}-GF_{254} (1 + 1) pyridine-benzene-acetonitrile-water (25 + 40 + 30 + 3), 1 x 10 cm, NS	Ninhydrine[15] chlorine-o-tolidine [R. 50][15]
Dicarbonicanhydrides	2	A SA	300 200	180 30	Kieselguhr G 25% impregnated with polyethyleneglycole 1000, diisopropylether-formic acid-water-PEG 1000 (90 + 7 + 3 + 2), 1 x 15 cm, CS	Bromocresolgreen 0.5% in ethanol [R. 31][15] Schweppes R. [R. 114][15]

Epoxy resin

TFG with 3–5 mg, T-region 50–450 °C, solvent chloroform ethyl acetate-formic acid (49 + 49 + 2), detection 0.5% bromocresolgreen in ethanol

Intensiv yellow zones on blue back hRf-values and zone mapping see Table 13

Dicarbonicanhydride-hardened epoxy resins

Blue, brown or yellow zones near the start TFG-repeat, T-region 100–450 °C, solvent toluene-acetone-ammonia 25% (80 + 20 + 1), detection: UV_{366} of sprayed zones with 0.02% fluorescamine in acetone

Variously coloured fluorescing TFG-zones, hRf-values and zone mapping see Table 12

With aromatic diamines or cycloaliphatic diamines hardened epoxy resins

Turquoise or non-fluorescing zones near the start

TFG-repeat, solvent pyridine-benzene-acetonitrile-water (25 + 40 + 30 + 3), detection of TFG-zones with chlorine-o-tolidine-reagent

Yellow zones with violet edge, resp. violet zones, hRf-values and zone mapping see Table 14

Dicyanodiamide-hardened epoxy resins

Yellow zones only near the start

Polyethylenpolyamine-hardened epoxy resins

Scheme 4. Scheme for group analysis of epoxy resin-hardeners by TLC on silica gel layers

V.2.4.2 Group Analysis of Hardened Epoxy Resins

To ascertain the group to which a hardener belongs, it is usually enough to separate and identify its products of thermal decomposition. This is best performed by TLC using suitable "guide solvents" in order to avoid interference and overlapping by the fragments of the starting materials which are transferred simultaneously; these then remain at the start or travel with the front. The compound classes expected as breakdown products of the hardeners can be detected by suitably selective reagents (cf. Table 10). Information about the hardener group is inevitably obtained when the chromatographic conditions, including the methods of detection used, are sensibly adapted to the separation problem and alternative analytical measures are carried out systematically. A suitably phased analytical scheme is given above for this.

The TFG- and TLC-data thus obtained on the hardener fragments studied are given in Tables 10 to 14.

V.2.4.3 Characterisation of Individual Hardeners of Hardened Epoxy Resins After Alkali Fusion

V.2.4.3.1 Preliminary Note

As has already been said, polyethylenepolyamine polyadducts give no thermal degradation products which contain amines. A variant of TFG was thus necessary

Table 11. TLC-conditions for polyalkylenpolyamines and TFG – quantities of hardened epoxy resins after alkaline fusion

	Layers, solvents, hRf-values		Evaluation			TFG-Start of Zones [°C]				b.p. [°C]
	I	II	Fluorescence in UV_{366} fluorescamine-reagent	Ninhydrine reagent[15]	Chlorine-o-Toli-dine-reagent[15]	DETA	TETA	PEHA	MBCA	
EDA	53–57	36–40	Turquoise	Red brown	Yellow	200	230	235	–	116
DETA	35–40	36–43	Turquoise	Red brown	Yellow	220	235	265	–	207
TETA	28–33	25–31	Turquoise	Red brown	Yellow	–	240	265	–	266 267
PEHA	10–20	5–20	Turquoise	Red brown	Yellow	–	–	265	–	–
Piperazine	30–35	5–10	Dark violet without fluorescence	Carmine	Yellow with Violet edges	–	230	225	–	146
MBCA	85–90	84–88 90–94	Intensive gray-green	Violet		–	–	–	230	–

I Silica gel HF_{254}-GF_{254} (1 + 1), ethanol-ammonia 25% (33 + 66), 1 x 10 cm, CS.
II Silica gel HF_{254}, acetone-ammonia 25% (85 + 15), 1 x 10 cm, CS.

Table 12. TLC-conditions for aromatic and cycloaliphatic amines and TFG-quantities of hardened epoxy resins (thermolysis T and alkaline fusion A)

	Layers, solvents, hRf-Values			Fluorescence quenching UV_{254}	Fluorescence UV_{366} fluorescamine-reagent[157]	Coloration ninhydrine reagent[15]	TFG-Start of Zones [°C]								b.p. [°C]
							DDM		DDS		m-PDA		MBCA		
	I	II	III				T	A	T	A	T	A	T	A	
Aniline	55–60	55–60	50–55	(+)	Yellow green	Burgundy	350	225	240	200	–	200	–	–	184
2-Toluidine	–	–	60–65	+	Light green	–	–	–	–	–	–	–	–	–	–
4-Toluidine	52–57	–	–	+	Ochre green	–	355	–	–	–	–	–	–	–	200 15 Torr
DDM	28–32	25–30	34–38	+	Orange brown	Purple	350	225	–	–	–	–	–	–	250
DDS	6–11	7–12	20–23	+	Turquoise	Burgundy	–	–	200	225	–	–	–	–	175 Fp.
m-PDA	15–20	14–19	25–30	+	Gray brown	Carmine	–	–	–	–	120	125	–	–	283
MBCA	0–5	–	–	–	Gray green	Violet	–	–	–	–	–	–	280	230	–

I Silica gel HF_{254}-GF_{254} (1 + 1), toluene-acetone-ammonia 25% (80 + 20 + 1), 1 x 10 cm, CS.
II Silica gel HF_{254}, toluene-acetone (85 + 15), 1 x 10 cm, CS.
III Silica gel HF_{254}, benzene-chloroform-methanol (60 + 30 + 10), 1 x 10 cm, CS.

Table 13. TLC-conditions for dicarbonic anhydrides and dicarbonic acids and TFG-quantities of hardened epoxy resins (thermolysis T and acidic desintegration after alkaline fusion SA)

	Layers, Solvents, hRf-Values		Detections: Fluorescence-quenching UV_{254}	Bromocresol-green-reagent[15]	Schweppes reagent glucose-aniline[15]	TFG-Start of Zones [°C]: MSA		PSA		HPA		MHPA		m.p. [°C]
	I	II				T	SA	T	SA	T	SA	T	SA	
MS	8–13	8–10	+	Yellow	Dark	100	200	–	–	–	–	–	–	135
MSA	65–70	18–20	+			140	150	–	–	–	–	–	–	55
FS	48–50	33–35	+	Zones	Brown	135	150	–	–	–	–	–	–	300 Subl.
PS	20–24	14–17	+		Zones	–	–	150	100	–	–	–	–	210 Decomp.
PSA	65–70	73–76	+	On		–		280	130	–	–	–	–	132
HPS	33–37	35–38	–		On	–	–	–	–	300	125	–	–	194 *cis* 229 *trans*
HPA	70–75	78–80	–	Blue	Light	–	–	–	–	380	150	–	–	33 *cis* 146 *trans*
MHPS	35–40	46–49	–		Back	–	–	–	–	–	–	265	120	–
MHPA	70–75	83–85	–	Back		–	–	–	–	–	–	330	200	fluid

I Silica gel GF_{254}-HF_{254} (1 + 1); chloroform-ethyl acetate-formic acid (49 + 49 + 2), 1 x 10 cm, CS.

II Kieselguhr G 25% with polyethylenglycole 1000 impregnated (with fluorescence indicator), diisopropylether-formic acid-water-polyethylenglycole 1000 (90 + 7 + 3 + 2) upper phase, 1 x 15 cm, CS.

Table 14. TLC – conditions of decomposition products of dicyanodiamide-hardened epoxy resins (thermolysis T and alkaline fusion A)

	Layers, solvents, hRf-Values		Detections		TFG – start of zones [°C]		m.p.[°C]
			Chlorine-tolidine-reagent[15]	Ninhydrine-reagent[15]	T	A	
	I	II					
Cyanoamide	80–85	90–93	Yellow with violet edge	–	150	50	42
Dicyanodiamide	50–60	73–78	Yellow with violet edge	Red brown	65, 160, 310	50	211
Melamine	15–20	35–40	Yellow with violet edge	–	175	225	354
Dicyanodiamidine	3–6	30–35	Yellow with violet edge	–	–	225	105

I Silica gel HF_{254}-GF_{254} (1 + 1), pyridine-benzene-acetonitrile-water (25 + 40 + 30 + 3), 1 x 10, NS.
II Silica gel HF_{254}-GF_{254} (1 + 1), acetonitrile-petroleum ether-carbon tetrachloride-tetrahydrofurane-water-formic acid (65 + 8 + 8 + 8 + 8 + 3), 1 x 10 cm, NS.

for their characterisation, namely, alkali fusion in the temperature gradient[130]. This accomplishes a selective and relatively mild hydrolysis of the amine-glycidyl ether linkages, followed by nucleophilic displacement of the corresponding polyethylene-polyamine. Table 11 contains the fission temperatures found in TFG-analysis.

This procedure is suitable also for separation of the aromatic amines and for fractionated liberation of the dicarboxylic acids after acidifying the product of alkali fusion. The experimental data thus obtained are in Tables 12 and 13. After ascertaining the best fusion temperature for the TFG-analysis by alkali fusion, isothermal alkali fission in the optimum temperature range suffices for simultaneous analysis of the various hardener types. Direct comparison of the hydrolysis products of different hardeners is made possible through point or band application of the transfer material. The TAS- and TLC-conditions required for this are given in Table 10. For purposes of illustration the thin-layer chromatograms of the hydrolysates of the hardeners used, after TAS-transfer, are given below.

V. 2.4.3.2 Aromatic Diamines

Aniline (zones 1, 4 and 13) can be regarded as an "indicator" for this group of hardeners. The additionally necessary distinction between the hardeners DDM (zone 2), 4,4'-diaminodiphenylsulfone (DDS, zone 16), m-phenylenediamine (m-PDA, zone 6) and 4,4'-methylene-bis(2-methylcyclohexylamine)(MBCA, zone 22) depends on the reliable detection of the regenerated molecules of hardener. They yield highly characteristic fluorescence colours in UV_{366} with the specific spray reagent for amines, fluorescamine[157]. The remaining TLC-zones, which have not been identified in more detail, can serve also for "finger print" characterisation.

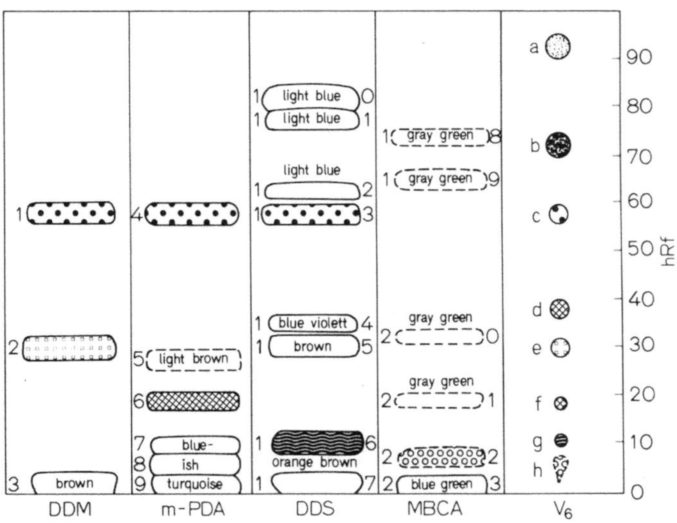

Fig. 17. TAS-transfer of aromatic and cycloaliphatic amines hardened bisphenol A epoxy resins from alkaline fusion. Experimental details see Tables 10 and 12. Reference mixture V_6: a) diphenylmethane, b) diphenylsulfone, c) aniline, d) 4.4'-bis[4-hydroxyphenyl]methane, e) 4.4'-diaminodiphenylmethane, f) m-phenylenediamine, g) 4.4'-diaminodiphenylsulfone, h) methylenbis[2-cyclohexylamine]

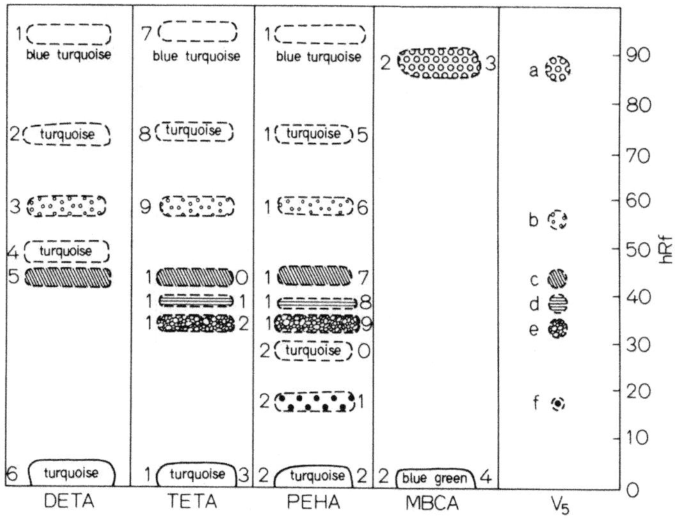

Fig. 18. TAS-transfer of polyethylenepolyamine hardened bisphenol A epoxy resins from alkaline fusion. Experimental details see Tables 10 and 11. Reference mixture V_5: a) methylenebis-[2-cyclohexylamine], b) ethylenediamine, c) diethylenetriamine, d) piperazine, e) triethylene-tetramine, f) pentaethylenehexamine

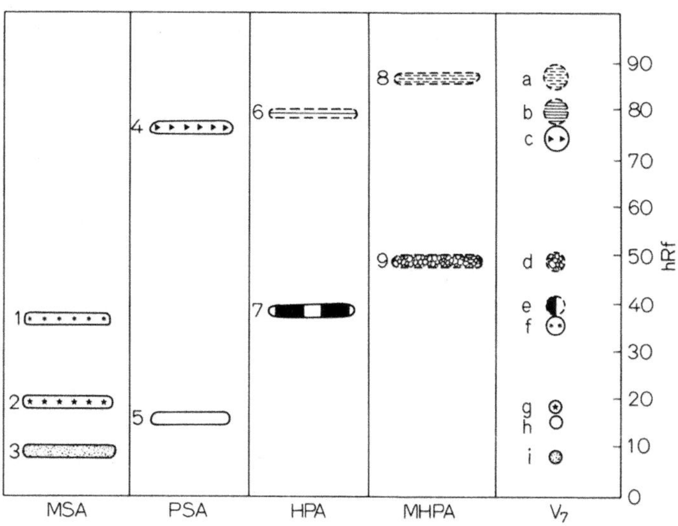

Fig. 19. TAS-transfer of dicarboxylic anhydride hardened bisphenol A epoxy resins from alkaline fusion after acidic desintegration. Experimental details see Tables 10 and 13. Reference mixture V_7: a) 3-methylhexahydrophthalic anhydride, b) hexahydrophthalic anhydride, c) phthalic anhydride, d) 3-methylhexahydrophthalic acid, e) hexahydrophthalic acid, f) fumaric acid, g) maleic anhydride, h) phthalic acid, i) maleic acid

V.2.4.3.3 Polyethylenepolyamines

Despite the procedural device used, identification of the polyethylene-polyamines presents problems. The technical materials used for hardening are usually mixtures of the declared hardener with its higher or lower homologues. It is nevertheless possible to detect the main components reliably. The declared higher polyethylenepolyamines always contain lower homologues but lower polyamines never contain higher homologues. The figures confirm this in an impressive way. The system hardened with diethylene triamine (DETA) shows the re-formed compound (zone 5) and also lower homologues, *e.g.* ethylenediamine (EDA, zone 3). In an analogous way, DETA (zone 10) and EDA (zone 9) appear with the type hardened with TETA (zone 12); piperazine is also found as an intramolecular cyclisation product (zone 11). The resin hardened with pentaethylenehexamine (PEHA, zone 21) shows consequently piperazine (zone 18) and the lower homologues, TETA (zone 19), DETA (zone 17) and EDA (zone 16).

TFG-analysis through alkali fission in the temperature gradient provides additional confirmation of the allocations with these critical classes of hardener. As a result of the differing stabilities of the particular polyadducts towards heat and hydrolysis in the alkali fusion, useful information is thus provided about the identity of their hydrolysis products yielded at different temperatures (cf. Table 11).

V.2.4.3.4 Anhydrides of Dicarboxylic Acids

Anhydrides of dicarboxylic acids can be obtained relatively easily as the dicarboxylic acids from the product of alkali fusion of the corresponding epoxy resins by acidifying with involatile mineral acids. They are transferred to the thin layer without difficulty at between 100 and 200 °C (cf. Table 13). Phenols, present as phenolates in the alkali mass and also freed from these by acidification begin to distil almost without exception at 50 °C[158]. They are quantitatively separated, permitting thermal fractionation of both substance classes. At the transfer temperature chosen (cf. Table 10), the carboxylic acids undergo partial dehydration to the original anhydrides. Maleic acid (MS, zone 3) yields maleic anhydride (MSA, zone 2) and also the thermodynamically more stable fumaric acid (FS, zone 1). Phthalic acid (PS, zone 5), hexahydrophthalic acid (HPS, zone 7) and 3-methylhexahydrophthalic acid (MHPS, zone 9) are accompanied in each case by the corresponding anhydrides (PSA, zone 4; HPA, zone 6; MHPA, zone 8).

V.2.5 Discussion and Summary

The TFG-analysis of epoxy resins depends on the defined thermal breakdown of the polyadducts to low molecular, mostly phenolic compounds, under conditions of thermolysis ($T_{max} < 500$ °C). The mechanisms of degradation which apply here have been extensively studied and accord with our analytical results[136, 137]. According to this, the thermal breakdown of epoxy resins begins with dehydration or dehydrogenation of the secondary alcohol function. This is followed by homolytic and/or

Hardened epoxy resin

Group analysis of hardeners by TFG with 3–5 mg, procedure see Scheme 4

Dicarbonic anhydrides	Aromatic diamines	Polyalkylenepolyamines	Dicyanodiamide

Isothermal alkali cleavage for the individual determination of hardeners
procedure see Table 10

MSA	m-PDA	DETA	DICY
PSA	DDM	TETA	Cyanoamide
HPA	DDS	PEHA	Melamine
MHPA	MBCA	(MBCA)	
(s. Table 13)	(s. Table 12)	(s. Table 11)	(s. Table 14)

Scheme 5. Identification scheme for the systematic hardener analysis with TFG

heterolytic bond fissions, depending on the individual bond stabilities. Further degradation products arise through rearrangements and cyclisations of intermediately formed compounds, capable of capture. According to our experience, the basic components of different polyepoxides differ strikingly in this. Such characteristic intermediates and their conversion products can be identified successfully in a simple and rapid way only by TFG. On the other hand, identification of the hardeners of thermostable epoxy resins with the help of thermal depolymerisation fails with the polyethylenepolyamines. This class of hardeners tends to undergo secondary thermal reactions and can therefore not be detected in the thermolysate. Hence the thermolysis of hardened epoxy resins through TFG is employed only for group analysis of the hardeners. Characterisation of individual hardeners is achieved by hydrolysis of the cross-linked polyglycidyl ethers through alkali fusion. This is a comparatively mild nucleophilic displacement of the amine or carboxyl group linked to the epoxy groups. Secondary thermal reactions are largely hindered because the alkali fission begins at fairly low temperatures. The fission reactions are thus standardised. This is shown by the reduction of fission products in TFG-analysis through alkali decomposition. This procedure variant can hence be recommended in principle for characterising individual hardeners and for simultaneous reference analyses (cf. Figs. 17, 18 and 19). The best analytical scheme is that given in Scheme 5 above.

V.3 Ester- and Amino-Condensation Polymers[118, 130]

V.3.1 Introduction[131–133]

Condensation polymers are derived from monomers which are at least bi-functional in a reaction of successive steps in which simple molecules (water, ammonia, HCl, alcohols) are eliminated. In the simplest case, macromolecules are formed by successive, intermolecular "self"-condensation of individual monomers containing two different condensable groups. The homopolycondensation products from ω-aminocarboxylic acids are an example. Industrially equally interesting are the combination products of two different monomers, each containing two identical reactive groups. This is exemplified by the polycondensation of dicarboxylic acids with diols to yield polyesters, of diamines and dicarboxylic acids to polyamines, and of diamines with bis-chloroformiates to polyurethanes. Linear polyamides, polyurethanes and polyester-polyurethanes, yielded according to other synthetic principles, are likewise treated here since they are related in structure and properties. Polyamides are easily formed by ionic homopolymerisation of cyclic amides (lactams) after ring fission. Polyurethanes and polyester-polyurethanes are yielded through polyaddition of di- and polyfunctional hydroxycompounds to di- and polyisocyanates. Polyurethanes are derived from di- and polyols. Polyester- or polyether oligomers with terminal alcohol function add to suitably reactive aromatic diisocyanates giving linear polyester-urethanes with prolonged chains. Since these contain terminal isocyanate groups, they can be hardened through cross-linking with water, glycols or diamines. As a result of their formation by successive reaction steps, condensation polymers are always mixtures of oligomers of regularly increasing molecular weight. Average molecule weights exceeding 20000–30000 are, however, seldom attained because this would require a degree of conversion of over 99%. Although its average molecular weight is low, this polymer class shows well-defined crystalline properties and is highly important as a starting material for qualitatively valuable products.

V.3.2 Analytical Procedures

Precise information about the chemical composition of technical condensation polymers is, for the reasons stated under 1., of appreciable scientific and industrial interest. IR-spectroscopy provides a first glimpse of their structure[5]. More detailed information, especially with regard to the monomers used, generally requires depolymerisation into products of lower molecular weight. This can be accomplished by thermal degradation or hydrolysis. The latter brings about a clearly defined "re-polymerisation". Pyrolysis-GLC and Pyrolysis-MS come primarily into consideration for the thermal fragmentation. Polyesters[55,57,66,80], polyamides[45,80,159], and polyurethanes[73] can be satisfactorily characterised. TLC is used to separate pyrolysates of polyesters, polyamides, and polyurethanes with parallel success but considerable advantage in manipulation and time[109]. Since the thermal breakdown of condensation polymers is relatively non-uniform, it is often replaced by acid or basic hydrolysis. This reverses the reaction of polycondensation and reforms the starting materials (re-polymerisation). Separation and detection are mostly performed with GLC. In this way, polyesters[2], polyamides[2,89,103] and polyurethanes[101] have been identified via the basic carboxylic acids, amines, and alcohols. These results have been obtained more simply and at the same time confirmed by TLC-analysis[2,89,101]. Alkali fusion in the pyrolysator of the GC-apparatus has brought an important methodic extension[104]. This "on-line" coupling can be regarded as a genuine step forward despite certain difficulties (see II.3.5.).

V.3.3 Thermofractography and Alkali Fission of Condensation Polymers[119, 130]

V.3.3.1 TLC-Conditions

A number of solvents proves suitable for TLC of the degradation products of condensation polymers. The Partridge mixture (upper phase of n-butanol-water-acetic acid, 40 + 50 + 10) separates adequately aliphatic amines from polyamides (cf. Table 16) (cf. Table 16). The aromatic diamines from polyols, lengthened by isocyanates, and ϵ-caprolactam can be separated surprisingly well with the more feebly polar solvent chloroform-ethanol (90 + 10) (cf. Table 16). Aliphatic polyols from polyesters and polyester-urethanes are chromatographed expediently with n-butanol, saturated with 25% ammonia (cf. Table 18). The separation of the lower diols is especially favoured by single development. Drying and a second development with the same solvent yields satisfactory hRf-distances between higher polyols, such as glycerol or pentaerythritol. The pertinent homologous aliphatic or isomeric aromatic dicarboxylic acids can sometimes be separated with surprising ease using chloroform – ethyl acetate-formic acid (49 + 49 + 2) on silica gel layers (cf. Table 20). Only for the separation of some crit-

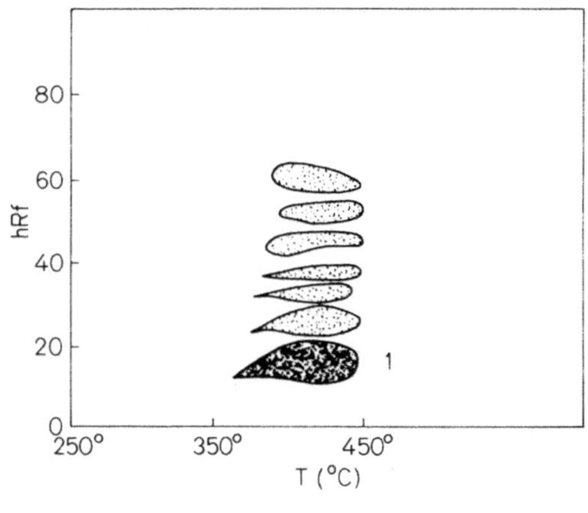

Fig. 20. TFG of 5 mg of polyamide 66. Characteristic for nylon types. Ninhydrinpositive zones are dotted. See Tables 15 and 16 for experimental details

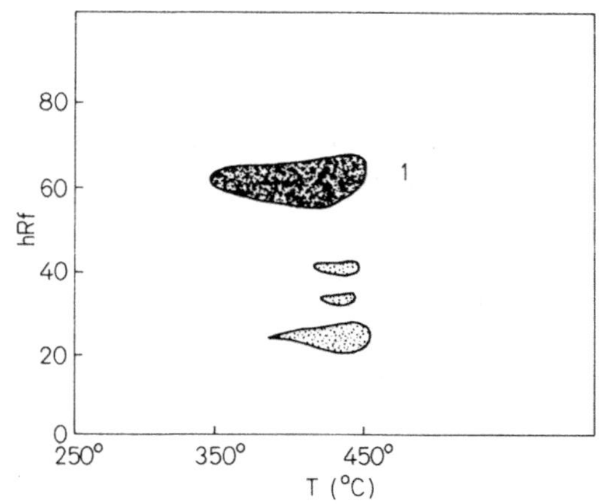

Fig. 21. TFG of 5 mg of polyamide 12. Characteristic for perlon types. Ninhydrinpositive zones are dotted. See Table 16 for experimental details

ical acid pairs (adipic/phthalic acid, sebacic/azelaic acid, cf. Table 20) is it best to resort to partition chromatography on kieselguhr layers, impregnated with 25% polyethyleneglycol 1000, using the upper phase of diisopropyl ether – formic acid – water – polyethylene glycol 1000 (90 + 7 + 3 + 2)[156]. Spray reagents, specific for various groups, are used to visualise the TFG-zones (cf. Tables 16, 18, 20). This substantially facilitates the identification of characteristic degradation products. For instance, with ninhydrin the aliphatic amines from linear polyamides always yield more or less red coloured zones whereas the aromatic amines from polyurethanes usually give other colours. Tests for the polyols, transferable sometimes from the alkaline fusion mass, sometimes after its acidification (cf. Tables 17 and 19), are the unspecific potassium permanganate-sodium metaperiodate reagent and, mainly, the selective colour reagents vanillin- or anisaldehyde-sulphuric acid (cf. Table 18). The dicarboxylic acids, separable from polyamides (cf. Table 15), from polyesters (Table 17) or from polyurethanes on a polyester basis (cf. Table 19) are sometimes detected through fluorescence quenching in UV_{254} or are visualised with various, although scarcely specific, indicator solutions (Reagents 31–34[15]).

V.3.3.2 Group Analysis of Condensation Polymers

Our studies of the degradation of condensation polymers using thermofractography have shown that the starting materials are not always reformed during thermal depolymerisation. For example, the expected dicarboxylic acids were not found in the TFG of polyamides. In contrast, the thermofractograms of the polyesters always showed the corresponding acid zones (cf. Fig. 22). Individual amino-condensation polymers could be distinguished from one another satisfactorily through characteristic TFG-zones (cf. Fig. 20–21). Hence, we employ TFG primarily for group analysis of condensation polymers, where it performs a useful service in the classification of polyamides into perlon- and nylon-types and their distinction from polyurethanes.

The essential classification features can be seen directly in the reproductions of thermofractograms of polyamide 66, a nylon type (cf. Fig. 20) and polyamide 12, a perlon type (cf. Fig. 21). Nylon types yield marked ninhydrin-positive TFG-zones

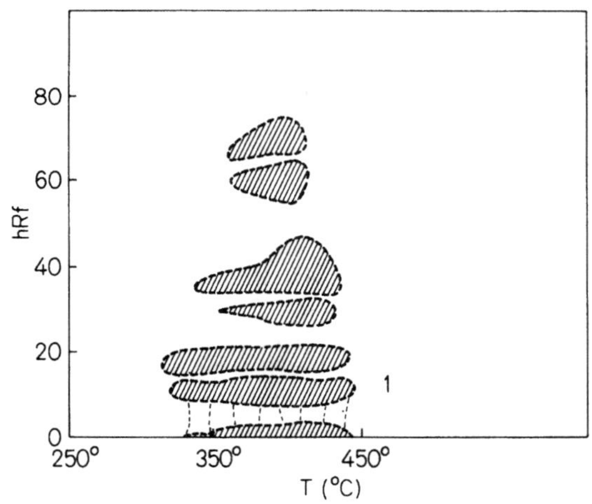

Fig. 22. TFG of 3 mg of polyethyleneterephthalate. In addition to the starting component, terephtalic acid (zone 1), other acid-positive zones showing fluorescence quenching in UV_{254} are observed over the whole hRf-region. See Tables 17, 18,20 for experimental details

in the lower hRf-region, from aliphatic diamines (zone 1, Fig. 20 = hexamethylene-diamine). Other zones, more feebly coloured, occur over the whole hRf-domain. In contrast, the most marked zones from perlon types are always at hRf-values near 60 (cf. Fig. 21, zone 1). These characteristic features adequately ensure their distinction, although both polyamide types begin to undergo fragmentation at ca. 350 °C. The typical red TFG-zones from detection with ninhydrin further distinguishes adequately both polyamide types from the corresponding polyurethane or polyester-urethanes which yield other colours (cf. Table 16). These last-named are characterised through their thermal breakdown which sets in already at 300 °C (cf. Fig. 23). Polyesters of the polyalkyleneterephthalate type can be extremely easily identified. As expected, they yield no ninhydrin-positive zones in the temperature range 300–450 °C. The acetic acid in the ninhydrin reagent is removed by briefly heating the layer and the acids can be visualised with the help of a suitable indicator (cf. Fig. 22).

The acids from the ninhydrinpositive polyurethanes on a polyester basis can be detected in the same way. On the other hand, these methods of detection do not work with the acid components from polyamides, as has been already mentioned elsewhere. However, the polyester-urethanes can be clearly distinguished from these through the variable start positions of TFG-zones (cf. Fig. 23) and the different colours with ninhydrin (cf. Table 16).

Based on the thermofractograms obtained from the group analysis, a simple analytical scheme for classification according to group of the condensation polymers studied here can be given. This is seen in Scheme 6 below.

V.3.3.3 Characterisation of Individual Condensation Polymers by Alkali Fission

The thermal depolymerisation of condensation polymers through thermofractography is excellently suited to group analysis of this polymer class but definitive identification of the particular type is guaranteed in only few cases. This is due to the lack of uniformity of the thermolysis reaction and the resulting variety of thermolysis products. Further, there is no thin-layer chromatographic system which

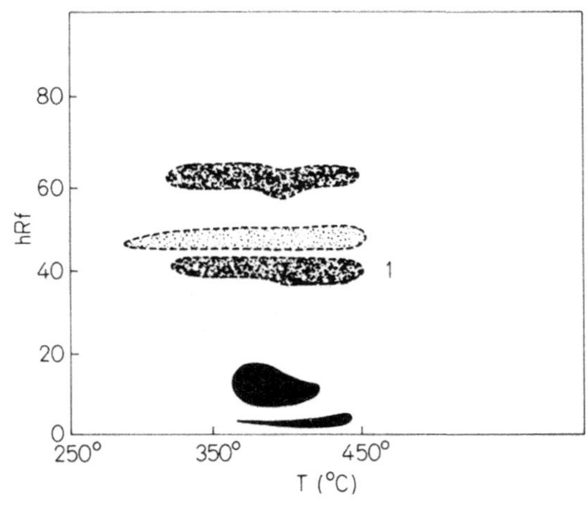

Fig. 23. TFG of 8 mg of Uperan 600 (polyester-urethane). The main component, 2,4-diaminotoluene, appears in the hRf-region 40–45 (zone 1). Other types normally differ from one another in the number, shape and hRf-range of the TFG-zones. See Tables 16 and 19 for experimental details

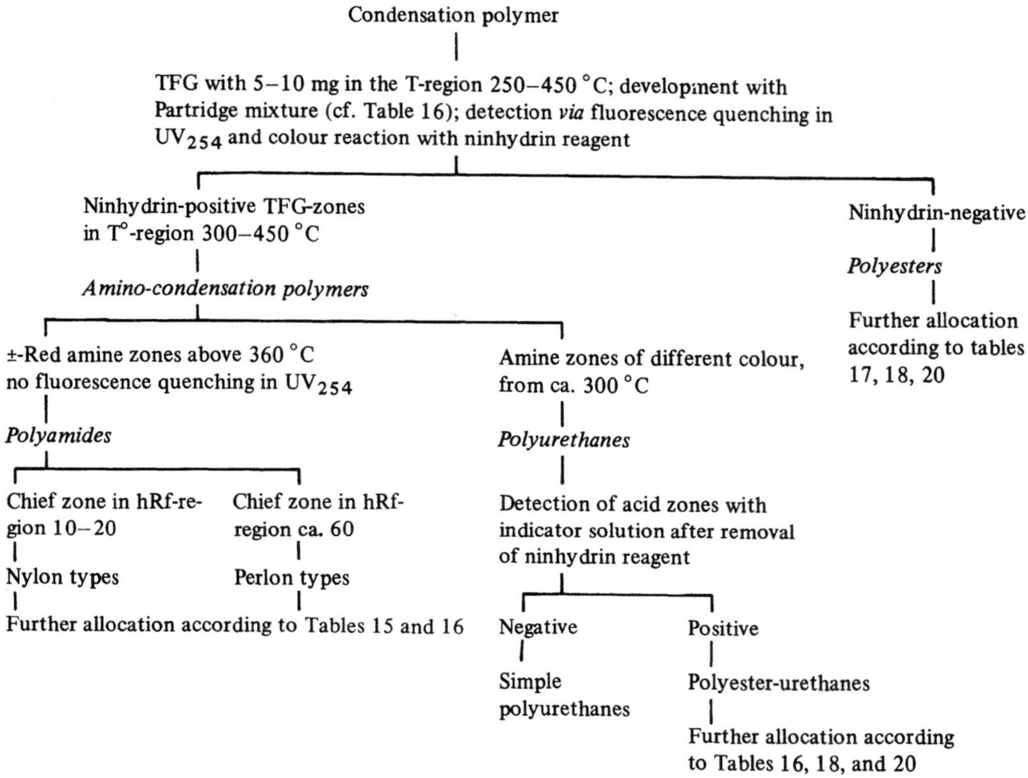

Scheme 6. Evaluation key for group analysis of condensation polymers using TFG

would enable all the products derived thermally from condensation polymers to be separated. It thus appeared advantageous to carry out a chemical prefractionation of the basic polyamines, neutral polyols and mono- and polybasic acids to be expected as breakdown products. Each compound class is then separately chromatographed and the characteristic chemical structural units selectively detected. The only procedure for this is isothermal alkali fission after having ascertained the best hydrolysis temperature through preliminary tests in the temperature gradient.

V.3.3.3.1 Polyamides

As known, polyamides are grouped into nylon types (A, B, C, Table 15) and perlon types (D, E, Table 15). Alkali fusion of the nylon types yields their structural bases of diamine and dicarboxylic acid. That of perlon types should lead to aminoacids but these undergo secondary thermal reactions under the temperature conditions obtaining here. Polyamide 6 is a special case since it unexpectedly re-forms ϵ-caprolactam during both thermal and hydrolytic breakdown, which enables it to be identified. The experimental conditions for the isothermal alkali fission of polyamides are given in Table 15.

Table 15. Analytic results through alkaline fusion of polyamides (weight 3 mg, oven temperature 300 °C, time of sojourn, 3–4 min, carrier gas 10 ml N_2/min)

Index	Polyamide type	To be transferred from alkaline fusion	To be transferred after neutralization
A	66	Hexamethylenediamine	Adipic acid
B	69	Hexamethylenediamine	Azelaic acid
C	610	Hexamethylenediamine	Sebacic acid
D	6	–	ε-Caprolactam
E	6 ATOM	Trimethylhexamethylene-diamine	Adipic acid and ε-Caprolactam

Table 16. TLC-conditions for polyamines and ε-caprolactam on silica gel HF_{254}-layer

Compounds	Solvents, hRf-values		Detections		
	Upper phase n-butanol – water – acetic acid (40 + 50 + 10) hRf	Chloroform – hRf	Fluorescence quenching UV_{254}	Ninhydrine reagent[15)]	Dragendorff reagent[15)]
Hexamethylenediamine	10–15	0	–	Dark red	–
Trimethylhexamethylenediamine	15–20	0	–	Dark red	–
4.4′-Diaminodiphenylmethane	95–100	60–65	+	Red	–
1.5-Diaminonaphthalene	95–100	70–75	+	Gray green	–
2.4-Diaminotoluene	95–100	40–45	+	Yellow	–
ε-Caprolactam	50	30–35	–	–	Yellow orange

Table 17. Analytic results through alkaline fusion of polyesters

Index	Polyester type	Weight [mg]	Temperature [°C]	Time of soj. [min]	To be transferred from alcaline fusion	To be transferred after acidifying
F	Polyethyleneterephthalate PETP	3	270	3	–	Ethyleneglycole and terephthalic acid
G	Modified PETP I	3	270	3	–	Ethylene glycole, adipic acid, and terephthalic acid
H	Modified PETP II	3	270	3	Neopentylglycole	Ethyleneglycole and terephthalic acid
I	Modified PETP III	3	270	3	–	Ethyleneglycole, azelaic acid, and terephthalic acid
J	Makrolon	1,5	250	3	–	Bisphenol A and CO_2
K	Hostadur	3	270	3	1.4-Butanediol	Terephthalic acid
L	Vestopal 120 (dispersion)	10	220	1	1.2-Propanediol	Fumaric acid, phthalic acid, and phthalic anhydride
M	Gu 6828 (dispersion)	10	220	1	–	Glycerol, phthalic acid, phthalic anhydride, and isononanoic acid

Details of the separation of the transferred polyamines and of ϵ-caprolactam are in Table 16 below. (see page *60*)

V.3.3.3.2 Polyesters

The degradation products of polyesters in alkali fusion are polyols and polybasic carboxylic acids; polycarbonates yield also diphenols. The optimum reaction temperatures vary and are compiled in Table 17.

It is noteworthy that certain polyols, *e.g.* glycerol and ethylene glycol, occur as salts in the alkaline mass and can be transferred only after acidification. They are completely transferred to the thin layer by 180 °C, so that they do not interfere with the dicarboxylic acids which pass over at higher temperatures (>260 °C); this provides an elegant separation of the two classes. The TLC of the hydrolysates of polyesters is dealt with in Tables 18 and 20.

Table 18. TLC-conditions for polyols on silica gel HF_{254}-layers

Compound	Solvent, hRf-value: n-Butanol, saturated with conc. ammonia — Single run	Double run	Detections: Na-metaperjodate K-permanganate reagent[15]	Vanilline-sulfuric acid[15]	Fluorescence quenching UV_{254}
Ethyleneglycol	40–45	60–65	Yellow	Greenish	–
Diethyleneglycol	50–55	65–70	Yellow	Greenish	–
1.2-Propyleneglycol	60–65	75–80	Yellow	Dark brown	–
1.4-Butanediol	80–85	85–90	Yellow	Gray black	–
Neopentylglycol	85–90	85–90	Yellow	Red violet	–
Glycerol	30–35	35–40	Yellow	Orange	–
Pentaerythritol	35–40	45–50	Yellow	Slight red	–
Bisphenol A	90–95	90–95	Yellow	Violet blue	+

Table 19. Analytic results through alkaline fusion of polyurethanes based on polyesters (weight 3 mg, temperature 260 °C, time of soj. 3 min, carrier gas 10 ml N_2/min)

Index	Polyester-polyurethane type	To be transferred from alkaline fusion	To be transferred after acidifying
N	Desmopan dd 1573	4.4′-Diaminodiphenylmethane and 1.4-Butanediol	Ethyleneglycol and adipic acid
O	Urepan 600	2.4-Diaminotoluene	Diethyleneglycol and adipic acid
P	Vulkollan 18	1.5-Diaminonaphthalene and little 1.4-butanediol	Ethyleneglycol and adipic acid

Table 20. TLC-conditions for organic acids

Compound	Layers, solvents, and hRf-values		Detection	
	Kieselguhr, impregnated with 25% polyethyleneglycole 1000 upper phase diisopropylether – water – formic acid – polyethyleneglycole 1000 (90 + 3 + 7 + 2)[156]	Silica gel HF_{254} chloroform – ethylacetate – formic acid (49 + 49 + 2)	Fluorescence quenching UV_{254}	Bromocresol-green reagent[15]
	hRf	hRf		
Adipic acid	40–45	30–35	–	yellow
Azelaic acid	80–85	55–60	–	zones
Fumaric acid	50–55	40–45	+	
Isononanic acid	100	70–75	–	on
Phthalic acid	15–20	30–35	+	
Phthalic anhydride	70–75	65–70	+	violet
Sebacic acid	85–90	55–60	–	
Terephthalic acid	60–65 tailing	0	+	back

V.3.3.3.3 Polyurethanes Based on Polyesters

The synthetic rubber products based on polyester-polyurethanes are broken down in a definite way by alkali fusion to polyamines, polyols and polycarboxylic acids. The best reaction temperatures for separating are 300 °C for polyamines, 180 °C for polyols and ca. 260 °C for polycarboxylic acids. Table 19 gives a survey.

The conditions in Tables 16, 18 and 20 have proved suitable for TLC-separation of the chemically pre-fractionated hydrolysis products which are thus yielded.

V.3.4 Discussion and Outlook

The thermofractograms of polycondensation compounds show that fragmentation occurs preferentially at particular positions of the polymer chain. This yields definite products in a reproducible way. Polyamides and polyurethanes suffer fission first at the relatively weak bond between carbonyl and imino groups, yielding the corresponding diamines. From the beginning of the zones of fission products it can be deduced that the bond at the isocyanate group in polyurethanes is easier to break than at the amide group. There are various ideas of the possible degradation mechanisms[160,161]. The thermolysis reaction is not uniform, so that many products are formed and some expected secondary products can be detected in the TFG. However, the results with polyester showed good agreement with the literature[162,163]. It is evident that TFG is excellently suitable for group analysis of condensation polymers (cf. Scheme 6). More definite degradation processes with more uniform secondary products are, of course, desirable to differentiate the various types within the same group of condensation polymer. This condition has been realised comparatively simply by reversing the reaction of formation of the polycondensation compounds with the help of a hydrolytic alkali fusion. The reaction mechanism corresponds to a so-called re-polymerisation. This means now that even small modifications (below 5%) of the polymer chains can be rapidly recognised with certainty.

V.4 Vinyl Polymers[130, 164]

V.4.1 Introduction[131–133]

Broadly speaking, vinyl polymers comprise all thermoplastic polymers derived from olefines and substituted olefines. They can be synthesised through free-radical homopolymerisation, with or without adding an initiator; or through ionic homopolymerisation, using Brönsted- or Lewis-acids or bases or organometallic initiators. Co- and Ter-polymerisation of polycomponent mixtures takes place by analogous mechanisms. Ethylenes containing real functional groups are used as starting monomers for the essential vinyl polymers. These must be distinguished from the polymers of homologous and isomeric alkenes which constitute an independent class of polymers, termed polyolefines. Well over a half of all the fully synthetic polymer materials are vinyl polymers. This accounts for the increased interest being shown in their rapid, certain and economic characterisation and in the analytical procedures developed for this.

V.4.2 Usual Methods for Analysis of Vinyl Polymers

Mostly gaseous products result from thermal degradation of the more technically important vinyl polymers, such as the thermoplastic polyethylene and its partly or completely halogensubstituted derivatives, *e.g.*, polyvinyl chloride; polyvinyl alcohol; polyacrylonitrile; polystyrene; synthetic rubbers based on hydrocarbons; and the co- and terpolymerisation products of these among themselves. They are thus generally analysed with the help of pyrolysis-gas chromatography or pyrolysis-mass spectrometry[1, 5, 8]. Pyrolysis-GC has proved to be excellent here, also because co- and terpolymer materials, *e.g.* ABS-terpolymers, yield characteristic pyrograms and even modified polymer components present in low concentration can be detected[8, 71, 45, 54, 62, 70, 64, 56]. Of the vinyl polymers which interest us (see Chapter V.4.3.), polyacrylates[46, 60, 62, 68, 167, 168], polyalkylacrylates[54, 56, 68, 165–167, 169] and polyvinyl carboxylates[46, 54, 56, 68, 165] have been subjected many times to pyrolysis-GC and the pyrograms thus obtained used for characterisation. Supplementary information can be obtained from direct IR-spectroscopic study of the vinyl polymers and IR-identification of their pyrolysates[5, 8]. However, there is as yet no fast and conclusive analytical method which permits selective characterisation of vinyl polymers within a reasonable time and with an acceptable outlay of apparatus. A first step in this direction was made with the TLC-analysis of test-tube-pyrolysates[26, 90] but the simple experimental equipment renders high and uncontrollable pyrolysis temperatures inevitable. The value of the procedure is reduced further by having to carry out transfers manually.

V.4.3 Thermofractography of Vinyl Polymers[130, 164)]

A prerequisite of TFG-analysis of polymers is the formation of volatile yet condensable products from their thermal degradation (cf. Chapter III.1). Its application to the identification of vinyl polymers is thus restricted to those types which yield, in addition to gases, less volatile products which can be adsorbed on a thin layer. For this reason, we have studied polyvinylesters, polyacrylates, polymethacrylates and the heterocyclic N-vinyl polymers, polyvinylcarbazole, and polyvinylpyrrolidone, which have been the objects of much acclaim recently. Gaseous degradation products of even the above mentioned polymer types can, however, be analysed without difficulty using a special TAS-GC-coupling device[128)].

V.4.3.1 TLC-Conditions

The vinyl polymers analysed here are polyolefines containing various substituents. It was therefore to be expected that the anticipated thermolysis products would not be separable on silica gel using a single solvent system. In all, 4 mobile phases were used for the thermolysis products, which sometimes showed rare fluctuations in polarity (cf. Table 21). As expected, the fatty acid components were formed in the TFG of polyvinyl esters (cf. Fig. 24 and Table 21). These can be separated relatively well using acetone-n-butanol-25%ammonia-water (65 + 20 + 10 + 5), at chamber saturation and in a run of 15 cm (cf. Fig. 24 and Table 21). Polyacrylates and polymethacrylates yield the corresponding monomers which can be separated satisfactorily with n-heptane-chloroform (50 + 50) (cf. Figs. 25 and 26). The fairly unpolar binary mixture of n-hexane-toluene (50 + 50) separates the breakdown products of polyvinylcarbazole (cf. Figs. 27 and 28, and Table 21). In contrast, the more polar fission products of polyvinylpyrrolidone need the more polar mobile phase of chloroform-ethylacetate (70 + 30) to obtain adequate hRf-differences. Reference may be made to Table 21 below for information about detection of the various thermolysis products.

V.4.3.2 Identification of Vinyl Polymers Using TFG

The thermofractography of vinyl polymers enables various groups of polymer to be distinguished and also individual types to be characterised. This is explained by the highly uniform mechanisms of thermolysis which always predominantly take the form of a retropolymerisation, irrespective of the type of polymer. The thermal breakdown of vinyl polymers is therefore very distinct and leads in a defined way to the corresponding monomers; sometimes these lose acetylene and form basic units which contain no vinyl group.

Table 21. TLC- and TFG-quantities for degradation products of vinyl polymers

Kind of polymer	weight [mg]	Degradation product	Begin of degradation °C	layers, solvents, hRf-values I	II	III	IV	Detection FQ UV_{254}	Methyl-red reag.[15]	Anisaldehyde sulfuric acid[15]	Dragendorff-reagent[15]	Iodine vapour[15]	Alkaline $KMnO_4$-solution[15]
Vinylacetate/vinyl-2-methyl-2-ethyl-heptanecarboxylate-copolymer	2	Acetic acid	290	28–33				–	+	–	–	–	–
		2-Methyl-2-ethyl-heptanoic acid	290	50–55				–	+	–	–	–	–
Vinylacetate/vinylpropionate-copolymer (1:1)	2	Propionic acid	300	35–40				–	+	–	–	–	–
		Acetic acid	290	28–33				–	+	–	–	–	–
Polybutylacrylate	0.5	Butylacrylate	325		38–40			+	–	–	–	+	+
Poly-n-butylmethacrylate/poly-ethylmethacrylate/polydecyl-methacrylate-mixture	1.5	n-Butylmethacrylate	200		42–48			+	–	–	–	+	+
		Ethylmethacrylate	195		35–40			+	–	–	–	+	+
		Decylmethacrylate	300		62–68			+	–	–	–	+	+
Polyvinylcarbazole	2.5	Carbazole	350			20–25		+	–	+	+	+	+
		Vinylcarbazole	420			70–75		+	–	+	+	+	+
		Dimer of vinylcarbazole	380			35–40		+	–	+	+	+	+
Polyvinylpyrrolidone	3	Pyrrolidone	260				5–10	–	–	+	+	+	+
		Vinylpyrrolidone	350				45–50	–	–	+	+	+	+

I Silica gel HF_{254}, acetone-n-butanol-ammonia 25%-water (65 + 20 + 10 + 5), CS, 1 x 15 cm.
II Silica gel HF_{254}, n-heptane-chloroform (50 + 50), CS, 1 x 15 cm.
II Silica gel HF_{254}, n-hexane-toluene (50 + 50), CS, 1 x 15 cm.
IV Silica gel HF_{254}, chloroform-ethylacetate (70 + 30), CS, 1 x 15 cm.

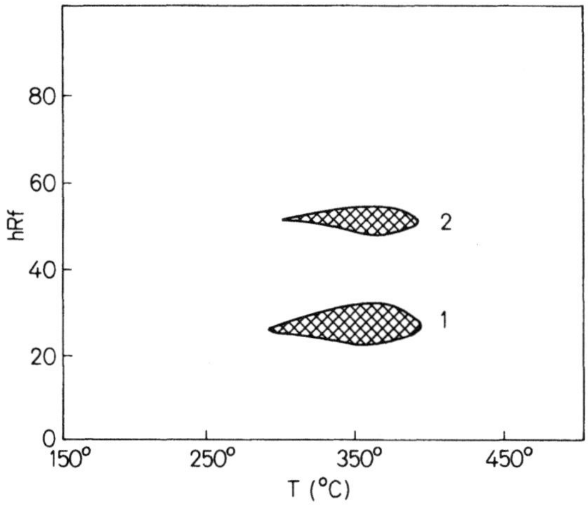

Fig. 24. TFG of the copolymer of vinylacetate and vinyl-2-methyl-2-ethylheptanecarboxylate. Irrespective of the type, the acid zones (1: acetic acid; 2: 2-methyl-2-ethylheptanecarboxylic acid) appear always in the region 300–400 °C and are detected as red zones on a pink background. See Table 21 for experimental details

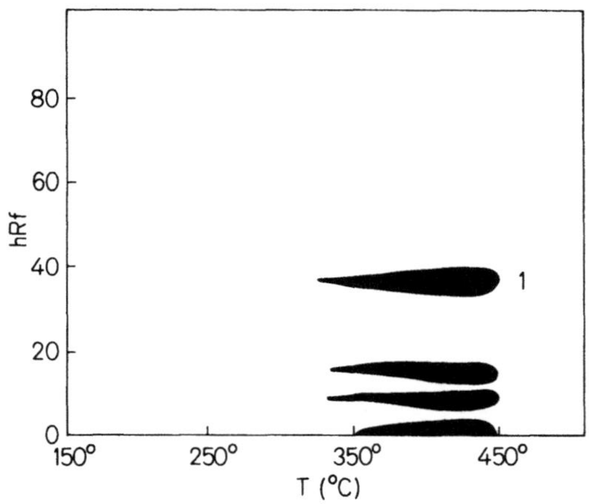

Fig. 25. TFG of polybutylacrylate. The commencement and shape of the ester zone (zone 1) of the monomer differ characteristically from those of corresponding methacrylates. Detection yield yellow zones on a violetred background. See Table 21 for experimental details

V.4.3.2.1 Polyvinyl Esters

The corresponding fatty acids always arise in the TFG of homo- or copolymers from vinylesters of these acids. The characterisation of the various vinylester types is based on their detection. Their fission begins at ca. 300 °C, whatever the fatty acid, and always takes place at the ester bond between carbon chain and carboxylic acid group. The TFG-conditions are mild (cf. Chapter III) and hence even long-chain carboxylic acids undergo no change during transfer. The copolymer of vinylacetate and vinyl-2-methyl-2-ethylheptanecarboxylate thus yields the TFG-products (see Fig. 24) acetic acid (zone 1) and 2-methyl-2-ethylheptanecarboxylic acid (zone 2). A 1 : 1-mixture of polyvinylacetate and polyvinylpropionate gives acetic and propionic acids. It is striking in both cases that acetic acid is still formed from polyvinylacetate even above 350 °C; this conflicts with the results of stepwise pyrolysis-GC[54]. It is further noteworthy that the sample required for TFG is only one third of that customary in pyrolysis-GC.

V.4.3.2.2 Polyacrylates

The commencement, shape and number of zones in the TFG of polybutylacrylate (see Fig. 25) are typical for this group of polyvinyl polymers. The higher fragmentation temperature compared to that of polybutylmethacrylate is evidently due to the greater cohesion between the individual polymer segments. Under the conditions used, small amounts of lower homologues are encountered along with the monomer itself (zone 1). Noteworthy also here is the small amount of sample needed. Polybutylmethacrylate in equally small amounts gives, in contrast, no detectable zones of monomer.

V.4.3.2.3 Polymethacrylates

As known, polymethacrylates split off only the corresponding monomers. This is reflected in the thermofractogram of a polymer mixture (see Fig. 26 below). There are, nevertheless, characteristic differences between individuals. n-Butyl- and ethylmethacrylates appear in the form of a double spindle (zones 2 and 1, respectively) whereas decylmethacrylate occurs as a single spindle form (zone 3). Since free monomers cannot be present in the polymer mixture, this finding demands the assumption of two competing fission mechanisms for poly-n-butyl- and polyethylmethacrylates. The chemical structure of the monomer evidently influences appreciably the degradation behaviour of polymethacrylates.

V.4.3.2.4 Polyvinylcarbazole and Polyvinylpyrrolidone

It can be seen from the thermofractograms that both polyvinylcarbazole and polyvinylpyrrolidone undergo thermal fragmentation to the corresponding monomers. In both cases, monomers minus the vinyl group can be detected. It was possible to prove that only the vinyl compound comes directly from the polyvinylcarbazole. The carbazole, detectable at above 350 °C (zone 1, Fig. 27), is formed from this vinylcarbazole (zone 3, Fig. 27) by an acid-catalysed, thermal transformation only on the silica gel layer. The reaction proceeds *via* a detectable intermediate (zone 2, Fig. 27), probably dimeric vinylcarbazole under elimination of acetylene. In contrast, vinylpyrrolidone (zone 2, Fig. 28) is thermally stable. This means that pyrrolidone,

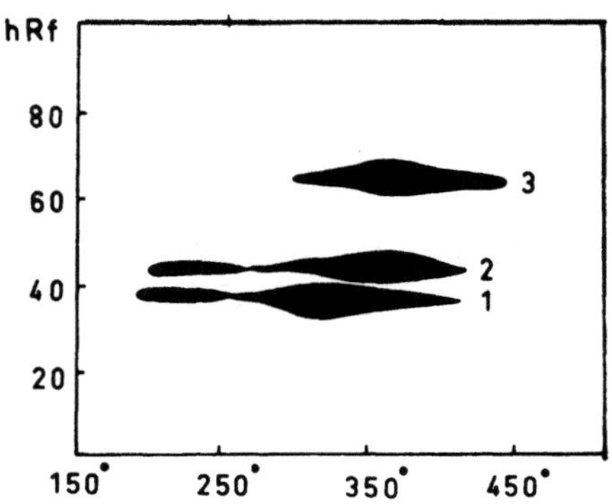

Fig. 26. TFG of a mixture of poly-n-butyl-, polyethyl- and polydecyl methacrylates. A double spindle beginning at ca. 200 °C occurs only with the first two. See Fig. 25 and Table 21 for detection and experimental details

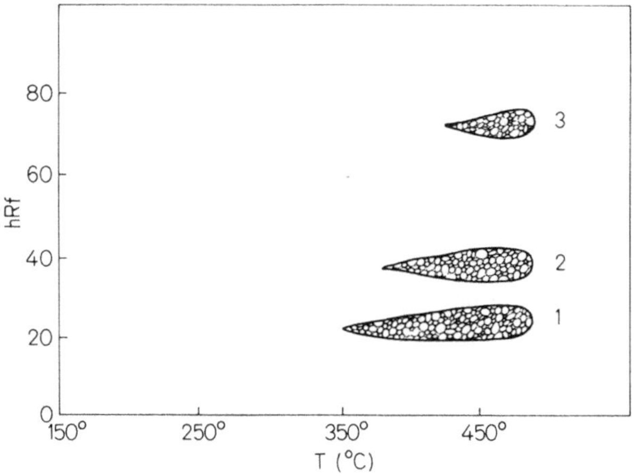

Fig. 27. TFG of polyvinylcarbazole, free of monomer. Carbazole (zone 1) is formed in a secondary reaction on the thin layer from vinylcarbazole (zone 3) via an intermediate (zone 2). Detection with anisaldehyde-H_2SO_4 yield dark blue zones. See Table 21 for experimental details

yielded at above 250 °C, comes directly from thermal breakdown of the polymer chain. The two vinylpolymers containing heterocyclic nitrogen which are considered here must thus presumably decompose according to different mechanisms; this is reinforced by the fact that polyvinylcarbazole begins to undergo thermolysis only at 350 °C. Incidentally, commercially available polyvinylcarbazole always contains the free monomer. This is easily seen in the TFG as a double spindle beginning at 150 °C.

V.4.4 Discussion of the Results

The thermofractography of vinyl polymers which yield condensable thermolysis products can be summarised as follows. All re-form the original monomer, with the

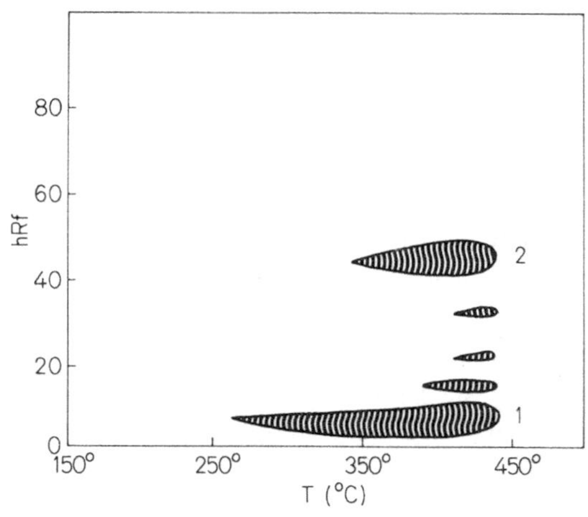

Fig. 28. TFG of polyvinylpyrrolidone. Pyrrolidone (zone 1) is found along with monomeric vinylpyrrolidone (zone 2). Detection with Dragendorff-reagent gives orange-brown zones. See Table 21 for experimental details

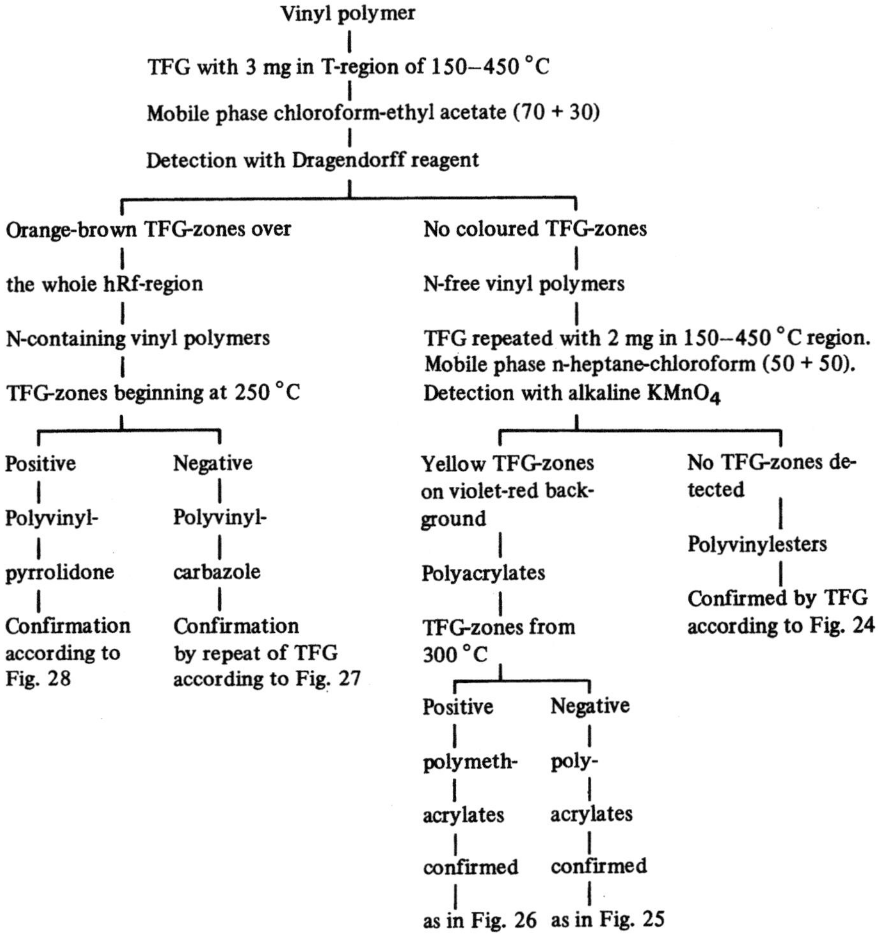

Scheme 7. Systematic identification of vinyl polymers

sole exception of the polyvinylesters. These last-named yield the corresponding carboxylic acids as a result of the unstable ester bond. This indicates the uniform thermal fragmentation reactions of vinyl polymers. Structural differences do impair this regularity unpredictably in some types of polymer but this helps to identify these. Thus polyacrylates undergo thermolysis with much greater difficulty than polymethacrylates since the links between the individual polymer units in the latter are appreciably loosened through the influence of the hyperconjugation of the methyl groups. Within the polymethacrylate group there are striking variations in the pattern of TFG-zones, essentially due to the nature of the monomer. It remains to say that the thermofractograms obtained permit unequivocal identification of the individual vinyl polymer types. An analytical scheme adapted to the TFG-method (Scheme 7) offers a systematic attack on the analytical problem.

VI Thermofractography of Polymer-Additives

VI.1 Separation and Identification of Plasticisers from PVC[170)]

VI.1.1 Preliminary Remarks[11, 171, 176)]

Through its manifold favourable properties and possibilities for fashioning polyvinyl chloride (PVC) deservedly enjoys the special regard of the plastics engineer. This is illustrated clearly by its dominant place in the registered production capacities of mass-produced thermoplastic materials. The broad field of application of PVC is, however, primarily due to certain additives, known as plasticisers. Thanks to these, the desired engineering physical properties of PVC-mixtures can be attained with defined accuracy over a wide domain. This modification of PVC, hard and brittle in its pure state, to a tough yet soft elastic condition, is termed "plasticising". A strict distinction must be made here between so-called "internal" and "external" plasticisers. The former act through copolymerisation of PVC with suitable plasticising copolymers, such as polyvinylesters or epoxy resins. The latter are worked into the polymer end product. A certain "solubilising power" or, at least, swelling capacity of the additive, mostly used in amounts from 20 to 120% of the polymer and usually of low molecular weight although sometimes also a polymer, is an absolute condition. As far as the quality of the plasticised PVC is concerned, however, the "external" plasticisers must be thermodynamically poor solvents. Nevertheless, they should have a sufficiently high affinity for the polymer to prevent separation during preparation or machining. Experience has shown so far that the "external" procedure still retains first place in the preparation of plasticised PVC. Primary plasticisers, miscible in all proportions with PVC, are distinguished from secondary plasticisers which are compatible with PVC only up to about 30%. Extenders, used only in mixtures with primary plasticisers, play a certain economic part. Another classification of plasticiser types is according to their mechanism of action. "Hinge plasticisers", *e.g.* the trialkyl phosphates, increase the mobility of the chain segments by interaction with the PVC-dipoles. In contrast, "umbrella plasticisers" derive their effect from the increase in distance between the chains in the vicinity of the PVC-dipoles by the long groups, as for instance in esters of aliphatic dicarboxylic acids. Others, such as the phthalates, are transition types. The connections and relations between the material properties of PVC and the plasticisers dealt with in this article (only external) are responsible for the high number and variety of the products of low molecular weight which come into consideration. Those of chief industrial significance are esters of aromatic and aliphatic dicarboxylic acids, such as phthalic, adipic, azelaic and sebacic acids, with monohydroxylic alcohols. They influence favourably the gelling power, stability, flexibility and insulating properties of the PVC-compounds concerned. In addition, trialkyl phosphates, citrates and trimellitates, and alkylarylsulfonates have aroused a certain interest in special applications. Polymeric plasticisers, mostly polyesters from aliphatic dicarboxylic acids with bifunctional alcohols, merit special attention on account of their low volatility and resistance towards oil.

VI.1.2 Hitherto Existing Procedures for Analysis of Plasticisers

The importance of the analysis of plasticisers is evident from the fact that they form an integral part of certain plastics. No adequate analytical characterisation of polymer systems is possible without determining the plasticisers. Since they have an appreciable influence on the properties of the host polymer, it is important to characterise them rapidly, reliably, and economically. To ascertain and compare physical data on plasticised polymers has little sense since the plasticisers influence these values in an unpredictable way. To subject them to analytical attack they must first be separated from the polymer. Usually mixtures of plasticisers are used and therefore the analysis itself consists of separation into the components by chemical or chromatographic means and then their chemical or spectroscopic identification. At present, there are two procedures, in principle the same, for separating plasticisers from host polymers; they depend on solubility differences between the two[11)]. In one procedure, the plasticisers are extracted with solvents in which the plastic basic material is insoluble. The other involves precipitation of the polymer from solutions of the plasticised product by means of solvents which selectively dissolve the plasticisers. The solutions must generally be concentrated before the final separation of the mixture of plasticisers. Gas and thin-layer chromatographic methods come preferably into consideration for this separation[1, 5, 11, 15)]. Less important is direct IR-spectroscopic characterisation of plasticisers without preceding separation, and of PVC-extracts[5, 11, 177, 178)]; and chemical detection by means of simple spot tests[11, 179)]. On the other hand, the HPLC-method, with its high selectivity, must be regarded as a welcome supplement and extension[14)]. However, chromatography is often unable to separate sufficiently the individual types of plasticiser. One may then recommend hydrolysis with alcoholic potassium hydroxide, followed by GC- or TLC-identification of the acids and alcohols yielded[11, 180)]. All methods of plasticiser analysis which have so far been described, have, however, the shortcomings, sometimes serious, listed below:

1. Most procedures for separating plasticisers from the polymers are involved and time consuming and require large amounts of solvent and of sample (>1 g); their success is by no means sure.
2. The chromatographic characterisation of plasticiser mixtures has so far been based on a single physical parameter (retention time R_t or hRf-value), resulting therefore in frequent confusion.

VI.1.3 TFG-Analysis of PVC-Plasticisers[170)]

In the light of our successful experiments in the thermal extraction of drug components from natural polymeric material[181)] and of flavouring and aroma products or dyes from foodstuffs and commodities[84, 85, 105–108, 114, 126)], it appeared advantageous to apply thermofractography, which had already shown its value in the analysis of natural and synthetic polymers (cf. Chapters IV and V), also to the analysis of polymer additives – in this case, PVC-ester plasticisers. Exploratory investigations showed that plasticisers can be easily separated from PVC and characterised by TLC, using TFG. A group analysis is also possible in principle *via* hydrolysis of the plasticisers through isothermal alkali fission in the TFG-apparatus and then determination of the acids and alcohols (cf. Tables 22 and 23 and Scheme 8). To this end, the fundamentals were worked out through TFG-analyses of uniform PVC-plasticiser mixtures and characteristic TFG-data of the individual plasticiser types were obtained (cf. Table 25). We limited our analyses to the class of ester plasticisers and three types of polymer species in order to keep a clear picture of our programme of study. The compounds involved are those given in Tables 24 and 25, and polypropylene adipate, polybutylene adipate and a polyphthalate, esterified with an acetylated polyol.

VI.1.3.1 Group Analysis of PVC-Plasticisers

It is impossible – which has been shown[11] – to separate all ester plasticisers using the same thin-layer chromatographic conditions. A group analysis has thus to be carried out to ascertain the individual ester types. For this, it is an advantage to obtain the acids by hydrolysis through isothermal alkali fission. It is advisable to confirm the group classification thus derived by TAS-application of the unchanged plasticisers followed by TLC-separation using a single mobile phase and specific group detection reagents. A preliminary hint about the identity of the plasticiser can be derived by identifying the alcohol components yielded by the isothermal alkali fission.

VI.1.3.1.1 Isothermal Alkali Fission of Ester-Plasticisers and TLC-Conditions for Separating the Acids and Alcohols Yielded

A selection of plasticiser types from each of the ester groups studied was submitted to isothermal alkali fission. The volatile alcohols and diols from the alkali fusion were chromatographed on silica gel using as mobile phase, n-butanol saturated with conc. ammonia and a run of 14 cm at chamber saturation. The results are presented in Table 22 below.

After transfer of the alcohol components, the carboxylic acids can be liberated from the alkali mass by acidifying with o-phosphoric acid. They can then be transferred under the same TAS-conditions. Table 23 contains the results obtained.

Group analysis of the ester-plasticisers can also be performed by thermal fission of the relevant types into the alcohol and acid components under the same conditions as in the alkali fusion but without alkali. A sample of about 0.5 to 1 mg then suffices. The thermolysis of the plasticisers takes place in an uncontrolled way and, for example, trimellitic acid cannot be detected with trimellitate samples.

VI.1.3.1.2 Group Analysis of Ester-Plasticisers Through TAS-Transfer

It is evident from the previous section that trialkyl phosphates and sulfonate-plasticisers escape group characterisation because their acid components are too involatile. TAS-separation of the plasticisers from the polymer system in combination with a TLC-group separation in the same solvent, dichloromethane, remedies

Table 22. TLC-conditions for alcohols and diols, transferred by TAS-Method from an alkaline fusion of plasticized PVC [weight 1–2 mg, oven temperature 300 °C, time of soj. 30 sec, carrier gas 10 ml N_2/min]

Short index	Type of plasticizer	Kind of alcohols	Layers, solvent, hRf-values, Silica gel HF_{254}-GF_{254} (1 + 1), n-butanol sat. with 25% ammonia 1 x 14 cm, CS	Detection[15] anisaldehyde-sulfuric acid
DOP	Di-2-ethylhexylphthalate	2-Ethylhexanol	80–85	Violet
DOA	Di-2-ethylhexyladipate	2-Ethylhexanol	80–85	Violet
DOZ	Di-2-ethylhexylazelate	2-Ethylhexanol	80–85	Violet
DOS	Di-2-ethylhexylsebacate	2-Ethylhexanol	80–85	Violet
ODTM	Octyl-decyltrimellithate	Octanol-decanol	80–85	Violet
TKP	Trikresylphosphate	–	–	–
ATBC	O-acetyltributylcitrate	Octanol from impurities	80–85	Violet
	Phenylpentadecylsulfonate	Pentadecanol	80–85	Violet
	Polypropyleneadipate	Propanediole (1.2)	50–55	Gray-brown
	Polybutyleneadipate	Butanediole (1.4)	60–65)	Gray-green
	Acetylated polyalkylene-Phthalate	Acetylated polyole	78–80	Blue gray

Further investigations on TLC-separation and -detection of long-chain alkylcarbinols and polyols – in the meantime – have showed chloroform-acetone-ammonia 25% (10 + 80 + 10)[15] to be a better separating solvent as that mentioned above in Table 22. In addition, potassiumdichromate-sulfuric acid[4, 15] has been proved to give a more clear and sensitive detection for alcohols (light blue zones on yellow back) as the reagent used above. More details will be reported[170].

Table 23. TLC-conditions for carboxylic acids, transferred by TAS-method after acidifying of the alkaline fusion of plasticized PVC [weight 1–2 mg, oven temperature 300 °C, time of soj. 30 sec, carrier gas 10 ml N_2/min]

Short index	Plasticizer	Kind of acid	Layer, solvent, hRf-values silica gel HF_{254}-GF_{254} (1 + 1) ethanol-water-ammonia 25% (80 + 8 + 12)[155], 1 x 15 cm	Detection 0.5% bromo-cresolgreen in ethanol, reag. 31 in[15]
DOP	Di-2-ethylhexylphthalate	Phthalic acid	13–16	Yellow
		Pthtalic anhydride	45–48	
DOA	Di-2-ethylhexyladipate	Adipic acid	25–30	
DOZ	Di-2-ethylhexylazelate	Azelaic acid	33–38	Zones
DOS	Di-2-ethylhexylsebacat	Sebacic acid	40–45	
ODTM	Octyl-decyltrimellithate	Trimellithic acid	2–7	On
		Phthalic acid	13–16	
		Phthalic anhydride	45–48	
ATBC	O-acetyltributylcitrate	O-acetylcitric acid	43–46	Violet
		Citric acid	18–20	
PPA	Polypropyleneadipate	Adipic acid	25–30	
PBA	Polybutyleneadipate	Adipic acid	25–30	
PAP	Acetylated polyalkylenephthalate	Phthalic acid	13–16	Back
		Phthalic anhydride	45–48	

Table 24. TLC-conditions for ester plasticizers on group analysis by TAS-transfer [weight 0.5 mg, oven temperature 250 °C, time of soj. 30 sec, carrier gas 10 ml N_2/min)

Group	Short index	Type of plasticizer	Layer, solvent, hRf-values dichloromethane, 2 x 15 cm	Detection FQ UV_{254}	Coloration
Short chain phthalates	DMP	Dimethylphthalate	38–41	+	Yellow zones on light yellow back with resorcinol-sulfuric acid [R. 210[15]]
	DEP	Diethylphthalate	39–42	+	
	DBP	Dibutylphthalate	56–60	+	
	DIBP	Diisobutylphthalate	56–60	+	
	DMGP	Dimethylglycolephthalate	3–7	+	
	BBP	Benzylbutylphthalate	56–60	+	
Long chain phthalates	DBzP	Dibenzylphthalate	68–70	+	Yellow zones on light yellow back with resorcinol-sulfuric acid
	DOP	Di-2-ethylhexylphthalate	72–76	+	
	DIOP	Diisoctylphthalate	71–75	+	
	DINP	Diisononylphthalate	73–77	+	
	DIDP	Diisodecylphthalate	75–79	+	
	DITDP	Diisotridecylphthalate	76–80	+	
Adipates	DOA	Di-2-ethylhexyladipate	41–45	–	Blue-gray zones on dirty yellow back
	DINA	Diisononyladipate	41–45	–	
	DIDA	Diisodecyladipate	42–46	–	
	BOA	Benzyloctyladipate	41–48	+	

Polymeric types special aliphatic a. aromatic types	DOZ	Di-2-ethylhexylazelate	42–46		–	with molybdato-phosphoric acid 10% in ethanol after heating to 110 °C for 10 min, reagent 158[15]
	DOS	Di-2-ethylhexylsebacate	57–61		–	
	ODTM	Octyl-decyltrimellithate	58–61		+	
	PPA	Polypropyleneadipate	17–19	unidentified	+	
	PBA	Polybutyleneadipate	17–19	unidentified	+	
	PAP	Polyphthalate	17–19	unidentified; 27–30 phthalic anhydride	+	
Phosphates and others	TKP	Tricresylphosphate	55–58		(+)	Yellow zones on violet back with alkaline $KMnO_4$-solution after heating to 100 °C. R. 142[15]
	DPK	Diphenylcresylphosphate	43–49		(+)	
	DPO	Diphenyloctylphosphate	35–42		(+)	
	TOF	Trioctylphosphate	35–39		–	
	ATBC	o-Acetyltributylcitrate	27–33		–	
	PPDS	Phenylpentadecylsulfonate	25–39		+	

Table 25. TLC and TFG quantities for ester plasticizers from PVC-compounds with thermofractography (weight 0.5–1 mg, temperature region 50–300 °C respect. 150–400 °C, heating rate 4 °C/min, slide speed 0.28 cm/min, carrier gas 10 ml N_2/min)

Plasticizer group	Short index	Type of plasticizer	TFG quantities begin of TFG-zones [°C]	Boiling range at reduced pressure 10 Torr [°C]	Layers, solvents hRf-values						Detection	
					I	II	III	IV	V	VI	FQ UV_{254}	Coloration
Short chain phthalates	DMP	Dimethylphthalate	75–80	134–138	32–35						+	Yellow
	DEP	Diethylphthalate	100–105	175–182	41–45						+	zones
	DBP	Dibutylphthalate	125	180–186	52–55						+	on slight
	DIBP	Diisobutylphthalate	120	171–177	58–60						+	back
	DMGP	Dimethylglycolephthalate	150	193–206	5–8						+	with
	BBP	Benzylbutylphthalate	160	235–255	50–55						+	resor-
Long chain phthalates	DOP	Di-2-ethylhexylphthalate	165	230–233		22–28					+	cinol-
	DIOP	Diisooctylphthalate	160	–		18–22					+	sulfu-
	DINP	Diisononylphthalate	185	244–252		20–24					+	ric acid
	DIDP	Diisodecylphthalate	210	250–267		23–28					+	reagent
	DITOP	Diisotridecylphthalate	245	–		26–31					+	210 in[15]
Adipates	DOA	Di-2-ethylhexyladipate	150	240–250			30–35				–	Orange
	DINA	Diisononyladipate	167	250–260			26–31				–	zones
	DIDA	Diisodecyladipate	180	241–250			29–34				–	on yellow-
	BOA	Benzyloctyladipate	175	235–255			17–21				+	ish back
	DBA	Dibenzyladipate	190	–			10–13				+	w. resor-
												cinol-sul-
												furic acid
												r. 210[15]

Aliphatic and	DOZ	Di-2-ethylhexylazelate	210	–	20–25				–	Bluegray z. on yellow back after heating to 110 °C with molybdato-phosphoric acid (R 158[15])
aromatic	DOS	Di-2-ethylhexylsebacate	215	–		33–38			–	
special types	O/DTM	Octyl/decyltrimellithate	260	–	25–30				+	
Phosphates	TKP	Tricresylphosphate	240	285–290		55–60			(+)	Blue z. on white back w. 2.6-dichloro-quinone-chloroimide-Borax, r. 59 [15]. White z. on light blue back w. ammoniummolybdate-hydrazine[11])
	DPK	Diphenylcresylphosphate	165	270–280		43–46			(+)	
	DKP	Dicresylphenylphosphate (impurity in DPK)	190	–		48–51			(+)	
	DPO	Diphenyloctylphosphate	175	–		50–55			(+)	
	TPP	Triphenylphosphate (impurity in DPO)	235	–		65–70			(+)	
	TOF	Trioctylphosphate	180			34–40			–	
others	ATBC	O-acetyltributylcitrate	205	–			39–42			1% KOH in EtOH, after heating to 100 °C yellow zones
	PPDS	Phenylpentadecylsulfonate	210	200–230				65–70	+	Yellow zones on blue back with alkaline $KMnO_4$[15])

I Silica gel HF_{254}-GF_{254} (1 + 1), petrolether-diisopropylether (30 + 70), 1 x 15 cm, CS.
II Silica gel HF_{254}-GF_{254} (1 + 1), petrolether-diisopropylether (90 + 10), 1 x 15 cm, CS.
III Silica gel HF_{254}-GF_{254} (1 + 1), petrolether-diisopropylether (80 + 20), 1 x 15 cm, CS.
IV Silica gel HF_{254}-GF_{254} (1 + 1), petrolether-diisopropylether (50 + 50), 2 x 15 cm, CS.
V Silica gel HF_{254}-GF_{254} (1 + 1), petrolether-diisopropylether (70 + 30), 2 x 15 cm, CS.
VI Silica gel HF_{254}-GF_{254} (1 + 1), dichloromethane, 1 x 15 cm, CS.

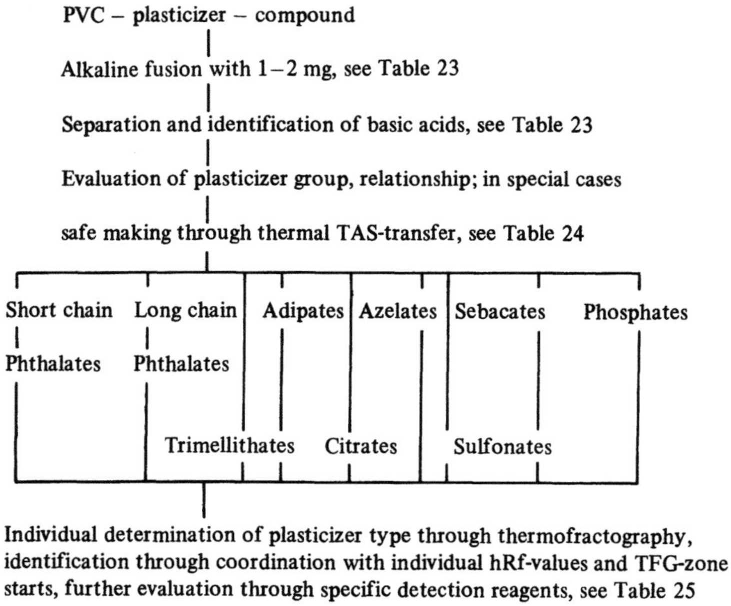

Scheme 8. Identification key for group analysis and individual determination of ester plasticizers through TFG

this deficiency, extends group classification of the various ester types to these missing phosphates and sulfonates, and at the same time establishes firmly the accuracy of the results of the alkali fission. These results (cf. Table 24) show clearly that in this way separation is possible into groups of related ester-plasticisers which appear together collectively in different hRf-regions. Further, detection by reagents for specific groups permits the individual types to be assigned to the same or different plasticiser classes. The results obtained in this procedure are given in Table 24.

VI.1.3.2 Characterisation of Individual Plasticisers Through Thermofractography

After having applied the results of alkali fission (cf. Tables 22 and 23) and TAS-transfer (cf. Table 24) to establish to which ester group a plasticiser belongs, the question arises of the identity of individual esters. This is a hard problem since the individual plasticisers within the same ester group can be separated thin-layer chromatographically only with difficulty or not at all[11]. A physical value in addition to the hRf-value of the particular ester is needed for its unequivocal identification. This additional information is obtained simply by thermal separation of the plasticisers from PVC-compounds in the temperature gradient through TFG. The programmed, linear increase of temperature fractionates the esters in each transition temperature range according to molecular weight. This relation is manifest since comparison of the commencements of TFG-zones with the boiling points of the plasticisers under reduced pressure shows the same tendency. This means that the thermofractography of distillable or sublimable compounds corresponds to distillation or sublimation under reduced pressure and can be regarded as an equally gentle fractionating procedure. This has been found also in analytical studies of homologous fatty acids[182]. Our results (cf. Table 25) lead us to consider the TFG-zone-commencements of ester platicisers as a supplementary analytical feature for characterising individual types, the hRf-values of which differ little from one another. This is the first analytical procedure which permits two-dimensional characterisation of the composition of plasticiser mixtures from PVC. We are sure that our TFG data on uniform PVC-plasticiser systems will enable the composition of technical mixtures of many components to be ascertained. This is the same of a future publication[170]. The data obtained are in Table 25 (see p. 78–79).

VI.1.4 Discussion and Summary

Thermofractography is suitable for the analytical characterisation of esterplasticisers from "soft-PVC". Isothermal alkali fission yields the acid components. Their identification serves for group classification of an unknown plasticiser (see Scheme 8). This findings can be fortified *via* TLC-separation of the plasticisers in the same mobile phase after thermal transfer and/or by detection with reagents for specific groups. When the group to which a plasticiser belongs is known, its TFG-zone commencement together with its own hRf-value give a relatively certain clue to its identity. Scheme 8 contains instructions for systematic plasticiser analysis with the help of TFG.

In our opinion, the essential advantages of our method are:
The extension of the analytical characterising factors of plasticisers from soft-PVC;
the elegant simplification of separation of the plasticisers from the plastic material by thermal fractionation;
and the improved procedure of direct coupling of separation and identification systems.

VII Conclusion

This retrospect of the investigations of polymers using thermofractography made so far by us shows that this method permits fast and easy analysis of macromolecular materials of any origin, provided that their thermal degradation products are volatile and then condensable on a thin layer. The thermofractograms thus obtained give a two-dimensional picture of the polymer composition. Each polymer type, natural or synthetic, yields its specific pattern of zones, characterised by their number, shape, commencement and hRf-region. Even where the zone-producing fragments have not been identified, the individual thermofractograms are characteristic of the particular polymer materials. The selectivity, specificity and reproducibility of the procedure as applied to polymer analysis demands comparison with dactyloscopy in criminology.

In the course of our investigations, many characteristic degradation products of every polymer class studied have been identified, and suitable chromatographic conditions worked out. This preliminary work now makes it possible by co-chromatographing authentic compounds to assign directly unknown polymers to their particular group and to determine their chemical nature. For this purpose, alternative schemes for identification are given for each polymer class. Further, clues to intermediates and reaction mechanisms of polymer degradation are revealed, which makes TFG interesting also for the theoretical chemists in polymer science.

Certain requirements of an analytical method for polymers are made on the industrial side: TFG fulfils many of these, *e.g.* simple handling, fast execution, practicability, easy mode of operation, small sample size (always less than 10 mg), modest outlay of apparatus, no great financial and personnel demands, uncomplicated interpretation of the analytical features. Summarising, it may be observed, that, respecting all the points mentioned above, it has been possible already to identify a series of polymer groups by means of TFG (cf. Table 26). Moreover, we believe that TFG-analysis has good prospects with other groups of polymers and additives given in Table 26.

Table 26. Application of TFG for the analyses of high polymers (classification by Hummel and Scholl[5])

Kind of material	Class of polymers	Types of compounds	Expected TFG-products
Natural products	Acidic resins and balms	Polycyclic aliphatic and acids and esters	Abietic acid and derivatives, phenols, essential oils
	Ester resins and balms	Cycloaliphatic and aromatic resin esters	Resin acids, -esters, -alcohols
	Resen resins and balms	Resin-esters, -acids, -alcohols, resens	Cycloaliphatic and aromatic-carbonic esters and alcohols
	Vegetable and animal oils and waxes	Mono-, di-, and triglycerides	Fatty acids, fatty esters and alcohols, polyenes, glycerol[182]
Biopoly-mers	Polysaccharides	Cyclic polyalcohols, -acids and esters	Sugars and derivatives[158]
	Polypeptides	Polycarbonicamides	Amino acids
	Lignins	Polycyclic phenylesters	Phenols, phenolalcohols[119] Phenolaldehydes[120, 158]
	Tannins	Gallotannins, ellagen, catechines, catechingallates	Phenols, phenolic acids[126]
Modified bio-polymers	Maleinat resins	Carbocyclic polycarbonican-hydrides	Maleic acid and anhydride, abietic acid, polyalcohols
	Phenol resin-modified natural resins	Natural resin phenylesters, albertol acids	Phenols, abietic acid and derivatives, chromanes
	Bitumina and tar, pitches	Condensed cycloaliphates and aromates	Polycyclic aliphates and aromates
Plastics	Phenol resins	Phenol-formaldehyde-condensates	Phenols[129, 130]

Kind of material	Class of polymers	Types of compounds	Expected TFG-products
	HC-resins	Cumarone-indene-polymerisates	Aromatic HC
	Vinylpolymers	Vinyl-heteroaromates and HC, vinylesters	Heteroaromates, HC, vinylesters[164], carbonic acids
	Polyesters	Polycarbonates, polyterephthalates, glyptal-resins, polyacrylates and -methacrylates	Phenols, polyols, carbonic acids[118] acrylates, methacrylates[164]
	Epoxy resins	Polyglycidylethers and -esters	Phenols, hardeners a. derivatives[136, 137]
	Aniline resins	Aniline-formaldehyde-condensates	Aniline and derivates
	Ketone resins	Cyclohexanone-formaldehyde-condensates	Cyclohexanone and derivatives
	Melamine resins	Melamine-formaldehyde-condensates	Melamine and methylolmelamines
	Polyamides	Dicarbonic acid-diamine-condensates, polyactams	Dicarbonic acids, diamines[118, 130], lactams
	Polyurethanes	Diisocyanate-polyol-adducts	Diamines, polyols[118, 130]
	Polysulfones	Bisphenol A-diphenolsulfon-polycondensates	Bisphenol A, diphenylsulfone
	Sulfonamide resins	Sulfonamide-formaldehyde-condensates	Sulfonamides and derivates
	Silicone resins	Aromatic polysiloxanes	Phenylsiloxanes
Plastic additives	Plasticizers	Carbonic- and phosphoric esters	Esters, acids, alcohols[108, 170]
	Organic pigments	Azo-, indigo-, anthraquinone-, phthalocyanine-triphenylmethane dye stuffs	Dye stuffs and derivatives
	UV-stabilizers	Benzophenones, salicylates, cyanacrylates, benzotriazoles	Free UV-stabilizers and degradation products
	Optical brighteners	Coumarin- and stilbene-derivatives	Free coumarins and stilbenes
	Antioxidants	Phenols, amines, chinolines	Free antioxidants
	PVC-stabilizers	Metal(Sn)-carboxylates and thiocarboxylates	Free stabilizers and derivatives
	Antistatica	Polyglycoles, polyglycole-ethers	
	Biocides	Triphenylmetallates, chlorinated HC	Me-organic compounds, free HC
	Vulcanization accelerators	Amino compounds, aromatic nitroso compounds, ureas, guanidines, dithiocarbamates, thiurames, imidazolines, piperidines, benzothiazoles	Free accelerators and degradation products
	Masticizers	Pentachlorothiophenols and others	Pentachlorothiophenol

Kind of material	Class of polymers	Types of compounds	Expected TFG-products
	Lubricants	HC, fatty alcohols and acids and esters, fatty amides, metallic soaps	Free compounds and degradation products
	Propellants	Azo- and hydrazo compounds, nitrosamines, thiotriazoles	N-free degradation products
	Thixotropics	Phenols, quinones, amines, chlorinated aromates, Ketoximes, aldoximes	Free compounds
	Adherents	Silanes	Free silanes

Acknowledgement. We thank the Deutsche Forschungsgemeinschaft for financial support and numerous firms for samples of materials.

VIII References

1. Hoffmann, M., Krömer, H., Kuhn, R.: Polymeranalytik. Stuttgart: Georg Thieme 1977, Bd. 1 und 2
2. Pohl, K. D., Bumiller, M. L.: Chem.-Ztg. *98 (7)*, 364 (1974)
3. Auterhoff, H., Klein, F. J., Rump, K. J.: PhiuZ *2 (3)*, 65 (1973)
4. Franck, R., Mühlschlegel, H. (Herausg.): Kunststoffe im Lebensmittelverkehr. Köln – Berlin – Bonn – München: Carl Heymanns 1974, Bd. 1 + 2, Lief. 1–19
5. Hummel, D., Scholl, F.: Atlas der Kunststoff-Analyse. München – Weinheim: Carl Hanser und Verlag Chemie 1968, Bd. 1
6. Krause, A., Lange, A.: Kunststoff-Bestimmungsmöglichkeiten. München: Carl Hanser, 1970, 2. Aufl.
7. Kraft, M.: Struktur und Absorptionsspektroskopie der Kunststoffe. Weinheim: Verlag Chemie 1973
8. Hummel, D., Trafara, G., Düssel, H. J., Lüderwald, J., Niedrig, H., Schulten, H. R., Rübenacher, K.: Pyrolytische Methoden in der Polymeranalytik: MS, GLC, IR-Spektroskopie, Anltg. z. GDCh-Fortb.kurs 87–78, Köln: Inst. f. Physik. Chemie der Univ. 1978
9. Krimm, S.: Adv. Polymer Sci. *2*, 51 (1960)
10. Hummel, D., Siegmund, E., in: Ullmanns Enzyklopädie der Technischen Chemie (W. Foerst Herausg.), 3. Aufl. München – Berlin: Urban + Schwarzenberg 1960, Bd. 11, S. 104–117
11. Wandel, M., Tengler, H., Ostromow, H.: Die Analyse von Weichmachern, in: Chemie, Physik und Technologie der Kunststoffe in Einzeldarstellungen (K. A. Wolf Herausg.). Berlin – Heidelberg – New York: Springer 1967, Bd. 11
12. Engelhardt, H.: Hochdruck-Flüssigkeitschromatographie. Berlin – Heidelberg – New York: Springer 1975
13. Dark, W. A., Conrad, E. C., Crossman, Jr., L. W.: J. Chromatogr. *91*, 247 (1974)
14. Groß, D., Strauß, K.: Kunststoffe *67 (8)*, 427 (1977)
15. Stahl, E. (Herausg.): Dünnschicht-Chromatographie. Ein Laboratoriumshandbuch, 2. Aufl., Berlin – Heidelberg – New York: Springer 1967
16. Inagaki, H.: Adv. Polymer Sci. *24*, 189 (1977)
17. Saechtling, H.: Kunststoff-Bestimmungstafel, 6. Aufl., München: Carl Hanser 1970
18. Saechtling-Zebrowski: Kunststoff-Taschenbuch, 19. Ausg. München: Carl Hanser 1974

19. Hellerich, W.: Kunststoffe – Eigenschaften und Prüfung. Stuttgart: Franckh 1968
20. El-Kodsi, G., Schurz, J.: Das Papier *27 (7)*, 253; *27 (10)*, 429 (1973
21. Kaiser, H.: Unverdorben, O.: Kunstst.-Berater *4*, 274 (1973)
22. Moustafa, A. B.: J. Oil Colour Chem. Assoc. *57 (1)*, 48 (1974)
23. Kupfer, W.: Z. Anal. Chem. *192*, 219 (1962)
24. Straschill, M.: Seifen-Öle-Fette-Wachse *95 (21)*, 815; *(22)*, 847; *(23)*, 874 (1969)
25. Braun, D.: Farbe + Lack *76 (7)*, 651–655 (1970)
26. Braun, D., Nixdorf, G.: Kunststoffe *62 (3)*, 187; *(4)*, 268; *(5)*, 318 (1972)
27. Taylor, L. J.: Chem. Tech. *9*, 552 (1973)
28. Wrasidlo, W.: Adv. Polymer Sci. *13, 1* (1974)
29. Mackenzie, R. C.: J. Macromol. Sci.-Chem. *A4 (4)*, 1015 (1970)
30. Rudloff, W. K.: NASA Report No. JJTRJ-C6221-9 (1971)
31. Farmer, R. W.: Thermochim. Act. *4*, 223 (1972)
32. Farmer, R. W.: Thermochim. Act. *4*, 203 (1972)
33. Miller, G. W.: Mat. Res. Stand. *12 (10)*, 8 (1972)
34. Sircar, A. K., Lamond, T. G.: Thermochim. Act. *7*, 287 (1973)
35. Wunderlich, B.: J. Thermal Anal. *5 (17)*, 117 (1973)
36. Gedemer, T. J.: J. Macromol. Sci.-Chem. *A 8 (1)*, 95 (1974)
37. Chiu, J.: J. Macromol. Sci.-Chem. *A 8 (1)*, 3 (1974)
38. MacCallum, J. R., Paterson, K., Richards, D. H.: Thermochim. Act. *9*, 79 (1974)
39. Sickfeld, J., Heinze, B.: Farbe und Lack *81 (2)*, 99 (1975)
40. Harms, D.: Anal. Chem. *25 (8)*, 1140 (1953)
41. Sugita, T., Ito, M.: Bull. Chem. Soc. (Japan) *38 (10)*, 1620 (1965)
42. Moskow, A. M., Emel'yanova, L. P., Karlinskii, L. E.: J. Appl. Spectrosc. (USSR) *10 (5)*, 860 (1969)
43. Cerceo, E.: Ind. Eng. Chem. Prod. Res. Develop., Polym. Sect. *9 (1)*,, 96 (1970)
44. Benco, G.: Tech. Ital. *36(4)*, 73 (1971)
45. Voigt, J.: Kunststoffe *51 (1)*, 18 (1961); *51 (6)*, 314 (1961)
46. Miller, D. L., Samsel, E. P., Cobler, J. G.: Anal. Chem. *33 (6)*, 677 (1961)
47. Fischer, W., Jaehn, L.: Adhäsion *9 (12)*, 487 (1965)
48. Levy; R. L.: Chromatogra. Rev. *8*, 48 (1966)
49. Fisher, G. E., Neerman, J. C.: I + EC, Prod. Res. Develop. *5 (3)*, 288 (1966)
50. Fischer, W., Meuser, H.: Gummi-Asbest-Kunststoffe *19*, 1229 (1966)
51. Perry, S. G.: J. Gas Chromatogr. *2*, 77 (1967)
52. Martinez, J., Guiochon, G.: J. Gas Chromatogr. *3*, 146 (1967)
53. Levy, E. J., Paul, D. G.: J. Gas Chromatogr. *3*, 1366 (1967)
54. Fischer, W. G.: G-I-T, Fachz. Lab. *11 (6)*, 562 (1967)
55. Davis, A., Golden, J. H.: J. Gas Chromatogr. *2*, 81 (1967)
56. Feuerberg, H.: G-I-T, Fachz. Lab. *13 (11)*, 1185 (1969)
57. Tsuge, S., Okumoto, T., Sugimura, Y., Takeuchi, T.: J. Chromatogr. Sci. *7 (4)*, 253 (1969)
58. Mosimann, H., Weber, W.: Schweiz. Arch. *36 (12)*, 402 (1970)
59. Tarnowiecki, H. M.: Mitt. Österreich. Kunstst.inst. *24*, 296 (1970); *25*, 85 (1971)
60. Bortniak, J.-P., Brown, S. E., Sild, E. H.: J. Forens. Sci., Techn. Sect. *16 (3)*, 380 (1971)
61. Jackson, M. T., Walker, J. Q.: Anal. Chem. *43 (1)*, 74 (1971)
62. Parrish, J. R.: Anal. Chem. *45 (9)*, 1659 (1973)
63. Levantovskaya, J. J., Dralyuk, G. V., Mochalova, O. A., Yurkova, J. A., Akutin, M. S., Kovarskaya, B. M.: Vysokomol. soyed. *A 13 (1)*, 8 (1971)
64. Alexeva, K. V., Khramova, L. P.: J. Chromatogr. *69*, 65 (1972)
65. Juvet, R. A., Smith, J. L. S., Li, K.-P.: Anal. Chem. *44 (1)*, 49 (1972)
66. Ball, G. L., Boettner, E. A.: J. Appl. Polym. Sci. *16*, 855 (1972)
67. Cascaval, C. N., Schneider, J. A., Rev. Roum. Chim. *17 (5)*, 835 (1972)
68. Roux, F. X., Audebert, R.: Analusis *2 (5)*, 343 (1973)
69. Luke, B. G.: J. Chromatogr. Sci. *11 (8)*, 435 (1973)
70. Braun, D., Canji, E.: Gummi-Asbest-Kunstst. *4*, 272 (1974)
71. Derby, J. V., Freedman, R. W.: Internat. Lab. *78*, 11 (1974)
72. Iglauer, N., Bentley, F. F.: J. Chromatogr. Sci. *12 (1)*, 23 (1974)

73. Burns, D. T., Johnson, E. W., Mills, R. F.: J. Chromatogr. *105*, 43 (1975)
74. Waysman, C., Matelin, D., Duc, C. L.: J. Chromatogr. *118*, 115 (1976)
75. Hummel, D. O., Duessel, H. J., Ruebenacker, K., Schweren, Th.: Makromol. Chem. *145*, 259, 267 (1971)
76. Sakamoto, R., Takahashi, Y., Ozawa, T.: J. Appl. Polym. Sci. *16*, 1047 (1972)
77. Andreev, A. P., Zaitseva, V. V., Anachenko, S. O.: Soviet Plast. *9*, 56 (1973)
78. Platz, K. H.: CZ-Chem.-Techn., *2 (8)*, 329 (1973)
79. Lüderwald, J., Ringsdorf, H.: Angew. Makromol. Chem. *29/30*, 444, 453 (1973)
80. Zeman, A.: Angew. Makromol. Chem. *416 (31)*, 1 (1973)
81. Hummel, D. O.: Polymer spectroscopy. Weinheim: Verlag Chemie 1974, p. 355
82. Sastre, R.: Inst. Plast. Cauch. Madrid (publ. in Rev. Plast. Mod.)
83. Merritt, Jr, C., Sacher, R. E., Petersen, B. A.: J. Chromatogr. *99*, 301 (1974)
84. Stahl, E., in: Analytical pyrolysis. Amsterdam: Elsevier 1977, p. 29
85. Stahl, E.: Acc. Chem. Res. *9*, 75 (1976)
86. Gedemer, T. J.: Soc. Plast. Eng. Am. Techn. Conf. Prep. *14 (5)*, 721 (1968)
87. Gedemer, T. J.: Soc. Plast. Eng. Am. Techn. Conf. Prep. *15 (5)*, 471 (1969)
88. Gedemer, T. J.: Plast. Des. Proc. *11*, 33 (1969)
89. Braun, D., Vohrendohre, G.: Kunststoffe *57*, 821 (1967)
90. Braun, D., Nixdorf, G.: Gummi-Asbest-Kunstst. *22*, 183 (1969)
91. Braun, D., Arndt, J.: Kunststoffe *62 (1)*, 41 (1972)
92. Braun, D., Lee, D. W.: Kunststoffe *62 (9)*, 571 (1972)
93. Mlejnek, D., Vorobjov, V.: J. Chromatogr. *79*, 91 (1973)
94. Meckel, L., Milster, H., Krause, U.: Textil-Prax. *16 (10)*, 1032 (1961)
95. Knappe, E., Rohdewald, J.: Z. Anal. Chem. *223*, 174 (1966)
96. Valk, G., Schliefer, K., Ruppert, K. H.: Textilindustr. *69 (3)*, 179 (1967)
97. Valk, G., Schliefer, K.: Textilindustr. *69(10)*, 783 (1967)
98. Schliefer, K.: Textil-Prax. *23 (10)*, 677 (1968)
99. Valk, G., Schliefer, K., Klippel, F.: Melliand Textilber. *49 (1)*, 92 (1968)
100. Schliefer, K., Valk, G.: Dtsch. Färberkalend. *75*, 517 (1971)
101. Braun, D., Mai, E.: Kunststoffe *58*, 637 (1968)
102. Braun, D., Jung, J. C.: Gummi-Asbest-Kunststoffe *23 (6)*, 618 (1970)
103. Tengler, H.: Plastverarbeit. *22 (5)*, 329 (1971)
104. Siggia, S., Frankosi, S. P.: Anal. Chem. *44*, 507, 2078 (1972)
105. Stahl, E.: J. Chromatogr. *37*, 99 (1968)
106. Stahl, E., Fuchs, J.: Dtsch. Apoth. Ztg. *108*, 1227 (1968)
107. Stahl, E.: Analyst. *94 (9)*, 723 (1969)
108. Stahl, E., Kraus, L. J., Groß, H.: Fette-Seifen-Anstrichmit. *73*, 357 (1971)
109. Patuska, G.: Gummi-Asbest-Kunststoffe *22 (7)*, 718 (1969)
110. Stahl, E.: Z. Anal. Chem. *261*, 11 (1972)
111. Brüderle, V.: G-I-T-Fachz. Lab., *21 (7)*, 649 (1977)
112. Stahl, E., Hartmann, L.: Z. Anal. Chem. *277*, 97 (1975)
113. Stahl, E., Becker, G., Brüderle, V.: J. Chromatogr. *129*, 41 (1976)
114. Stahl, E.: J. Chromatogr. *33*, 273 (1968)
115. Stahl, E., Müller, Th. K. B.: Z. Anal. Chem. *268*, 102 (1974)
116. Stahl, E., Müller, Th. K. B.: Z. Anal. Chem. *271*, 257 (1974)
117. Stahl, E., Müller, Th. K. B.: Z. Anal. Chem. *274*, 257 (1975)
118. Stahl, E., Oey, L. S.: Kunststoffe *64 (11)*, 657 (1974)
119. Stahl, E., Karig, F., Brögmann, U., Nimz, H., Becker, H.: Holzforschung *27 (3)*, 89 (1973)
120. Karig, F., Stahl, E.: Holzforschung *28 (6)*, 201 (1974)
121. Freudenberg, K. *et al.*: Ber. Dtsch. Chem. Ges. *73*, 167 (1940)
122. Creighton, R. H. J. *et al.*: J. Amer. Chem. Soc. *63*, 3049 (1941)
123. Hibbert, H.: Paper Trade J. *113*, 35 (1941)
124. Kratzl, K. *et al.*: Holz Rohst. Werkst. *23*, 237 (1965)
125. Björkman, A.: Svensk Papperstidn. *59*, 477 (1956)
126. Stahl, E., Karig, F.: Z. Anal. Chem. *265*, 81 (1973)

127. Freudenberg, K.: Handbuch der Pflanzenanalyse (Hrsg. G. Klein). Wien: Springer 1932. Bd. III/2, S. 1ff.
128. Stahl, E., Herting, T.: Chromatogr. *7*, 637 (1974)
129. Stahl, E., Oey, L. S.: Angew. Makromol. Chem. *44 (661)*, 107 (1975)
130. Oey, L. S.: Dissertation, Saarbrücken 1976
131. Stoeckhert, K.: Kunststoff-Lexikon, 5. Aufl. München: Carl Hanser 1973
132. Braun, D., Cherdron, H., Kern, W.: Praktikum der Makromolekularen Organischen Chemie, 2. Aufl., Heidelberg: Dr. Alfred Hüthig Verlag 1971
133. Vollmert, B.: Polymer Chemistry. Berlin Heidelberg New York: Springer 1973
134. Finn, F. N., Megson, N. J. L., Whitacker, E. J. W.: Chem. Ind. (London) *28*, 849 (1950)
135. Wegler, R., Regel, E.: Makromol. Chem. *9*, 1 (1953)
136. Stahl, E., Brüderle, V.: Angew. Makromol. Chem. 1978, in press
137. Brüderle, V., Stahl, E.: Colloid Polym. Sci. 1978, in press
138. Jahn, H., Goetzky, P., in: May, C. A., Tanaka, J., Epoxy resins – chemistry and technology, New York: Marcel Dekker 1973, pp. 653–696
139. Takeuchi, T., Kakugo, M.: J. Chem. Soc. Japan, Ind. Chem. Sect. *67*, 308 (1964)
140. Stuart, M., Smith, D. A.: J. App. Polym. Sci. *9*, 3195 (1965)
141. Keenan, M. A., Smith, D. A.: J. Appl. Polym. Sci. *11*, 1009 (1967)
142. Leisegang, E. C., Stephen, A. M., Patterson-Jones, J. C.: J. Appl. Polym. Sci. *14*, 1961 (1970)
143. Leisegang, E. C., Patterson-Jones, J. C., Stephen, A. M.: J. South Afr. Chem. Inst. *23*, 1 (1970)
144. Lee, L. H.: J. Appl. Polym. Sci. *9*, 1981 (1965)
145. Lee, L. H.: J. Polym. Sci. (Part A) *3*, 859 (1965)
146. Gac, N. A., Spokes, G. N., Benson, S. W.: J. Polym. Sci. (Part A 1) *8*, 593 (1970)
147. Vlastaras, A. S.: Vac. Microbal. Techn. *8*, 229 (1971)
148. Fleming, J.: J. Appl. Polym. Sci. *10*, 1813 (1966)
149. Sugita, T.: J. Polym. Sci. (Part C) *23*, 765 (1968)
150. Bishop, D. P., Smith, D. A.: J. Appl. Polym. Sci. *14*, 205 (1970)
151. Inagaki, M., Hayashi, M.: C. A. *73*, 4469q (1970)
152. Dacher, J. F., Guénier, J. P., Herve-Bazin, B., Moulut, D.: Chromatogr. *8 (5)* 228 (1975)
153. Parrish, J. R.: J. Chromatogr. *18*, 535 (1965)
154. Nascu, H., Hodisan, T., Liteanu, C.: Stud. Univers. Babes-Bolyai Ser. 1 (Chemia) *20*, 63 (1975)
155. Braun, D., Geenen, H.: J. Chromatogr. *7*, 56 (1962)
156. Knappe, E., Peteri, D.: Z. Anal. Chem. *188*, 184; 352 (1962)
157. Ranieri, R., McLaughlin, J.: J. Chromatogr. *111*, 234 (1975)
158. Karig, F.: Dissertation, Saarbrücken 1973
159. Feuerberg, H., Weigel, H.: Z. Analyt. Chemie *199*, 121 (1964)
160. Achhammer, B. G., Reinhart, R. W., Kline, G. M.: J. Appl. Chem. *1*, 301 (1951)
161. East, G. C., Kakar, S. K., Lupton, C. J., Truter, E. V.: Proc. Soc. Analyt. Chem. *10*, 93 (1973)
162. Allen, R. J. P., Forman, R. L., Ritchie, P. D.: J. Chem. Soc. (London) *3*, 2717 (1955)
163. Jengor, H. V. R., Ritchie, P. D.: J. Chem. Soc. (London) *4*, 2815 (1956)
164. Stahl, E., Oey, L. S.: Z. Anal. Chem. *275*, 187 (1975)
165. Barrall, E. M., Porter, R. S., Johnson, J. F.: Anal. Chem. *35*, 73 (1963)
166. Belenkii, B. G., Turkova, L. D., Andreeva, G. A.: Vysokomol. Soedin, Ser. B., *14*, 349 (1972)
167. Wallisch, K. L.: J. Appl. Polym. Sci. *18*, 203 (1974)
168. Bambough, K. J., Cook, C. E., Clampitt, B. H.: Anal. Chem. *35*, 1834 (1963)
169. Vacherot, M., Marchal, J., Hinschberger, A.: C. R. Acad. Sc., Ser. C. *264*, 962 (1967)
170. Stahl, E., Brüderle, V.: Kautschuk-Gummi-Kunststoffe, in preparation
171. Linhardt, F., Oschatz, F., in: Ullmanns Enzyklopädie der technischen Chemie (Hrsg. W. Foerst), 3. Aufl., München – Berlin: Urban und Schwarzenberg Bd. 18, S. 540ff.
172. Haase, H.: Chem. Ind. *27 (5)*, 275 (1975)
173. Kopsch, H.: Kunststoffe *67 (3)*, 141 (1977)

174. Stühlen, F., Meier, L.: Kunststoffe *61 (2)*, 867 (1971); *62 (10)*, 674 (1972)
175. Stühlen, F., Meier, L.: Kunststoff-Rundschau *19 (6)*, 251 (1952), *19 (7)*, 316 (1972)
176. Mair, H. J., Meier, L.: Kunststoffe *60 (5)*, 301 (1970)
177. Rath, H., Brielmaier, H.-J., Rau, J., Freyer, H.: Melliand Textilber. *42 (9)*, 1030 (1961)
178. Meise, W., Ostromow, H.: Kunststoffe *54 (4)*, 213 (1964)
179. British Standards Institution, BS 4835: April 1973
180. Braun, D., Vohrendohre, G.: Z. Analyt. Chem. *207*, 26 (1965)
181. Stahl, E. (edt.): Chromatographische und mikroskopische Analyse von Drogen; Eine praktische Ergänzung zu den Arzneibüchern Europas. Stuttgart: G. Fischer 1971
182. Werndorff, F.: Dissertation Saarbrücken 1977

Received May 10, 1978
H.-J. Cantow (editor)

Preparation and Properties of Star-branched Polymers

Stanley Bywater

Division of Chemistry, National Research Council of Canada, Ottawa, Canada K1A 0R9

The methods available for the preparation of star-branched polymers are described. After presentation of theoretical predictions of the effect of this type of branching on the properties of polymers, available experimental data are compared and critically discussed. Thermodynamic and visco-elastic properties at infinite dilution are first presented followed by visco-elastic properties in concentrated solution and the melt. Finally the effect of star-branching on T_g of low molecular weight polymers is described.

Table of Contents

1 Preparation of Star-branched Polymers

The preparation of polymers with well-defined forms of branching has posed a challenge to polymer chemists for many years. Branched polymers are often formed in polymerization reactions but it is difficult, in most cases, to define the number and types of branches and thus to correlate the changes in physical properties with molecular architecture. Star-branched polymers consisting of several linear chains linked together at one end of each chain constitute the simplest form of branching. While this type of branching is quite far removed from that found in commercial polymers, the study of star macromolecules of well-defined structure permits a more quantitative study of the effects of branching in one special form. It also provides an opportunity to check current theories of polymer solutions, for without changing the basic nature of the polymer under study, a change in segment density is produced on branching. The theories should be capable of dealing with this modification if they are to be generally successful.

Perhaps the earliest attempt to synthesize such molecules was that of Shaefgren and Flory[1] who polymerized ϵ-caprolactam in presence of either a tetrafunctional or an octafunctional carboxylic acid to produce polymers of the type $R\{-CO(NH(CH_2)_5CO)_n-OH\}_x$ where x = 4 or 8. The predicted narrowing of the molecular weight distribution from the most probable value ($\overline{M}_w/\overline{M}_n = 2$) found in linear polycondensates was observed. Melt and intrinsic viscosities were measured for polymers of $\overline{M}_w$ up to ~30,000. The eight-branch polymers were found to have slightly smaller (~20%) viscosities than their linear counterparts, little change being observed for the four-branch polymers. These results are however not typical, later investigations showing quite marked decreases in both types of viscosity.

Star-branched polysaccharides have also been prepared from glycogen and amylopectin. The addition of glucose-l-phosphate and the enzyme phosphorylase to maltotetrose or its higher oligomers produces linear amylose chains[2]. If glycogen or an amylopectin fraction is substituted for the maltotetrose, amylose side chains are grown from a number of points on the outer chains of the former polymers[3,4]. The nucleus in this case is already a quite large, relatively compact, branched macromolecule with a molecular weight of several millions. With glycogen some 15–20 amylose chains could be added of DP between 400 and 10,000. The outer chains of amylopectin appear to be more available to the enzyme which added about 10,000 side branches to form macromolecules of extremely high molecular weight.

Most of the star polymers described in the literature have been produced by anionic polymerization techniques. These provide, under suitable conditions, the ability to produce almost monodisperse linear polymer chains. If not deliberately terminated, moreover, the chain ends still possess highly reactive organo-alkali metal groups which can react to couple with multifunctional small molecules thereby producing regular star-branched macromolecules, of equal arm length. If the linking molecule is itself a polymer chain having a number of reactive sites, then a so-called comb polymer is formed having a number of branches of equal length attached to the backbone.

The techniques required to produce the precursor active polymer chains have been adequately described elsewhere[5–7]. Often these are formed by sec-butyllithium initiation of styrene, butadiene or isoprene. Polymerization proceeds to completion yielding a polymer of DP = M_0/I_0 after which the coupling agent can be added. The coupling agents most often used are chlorine containing compounds, the earliest examples being silicon tetrachloride[8] or chlormethylated benzene[9]. In the former case, for example, the following reaction should take place:

$$4 \sim\!\sim\!\sim^{-}Li^{+} + SiCl_4 \rightarrow (\sim\!\sim\!\sim)_4Si + 4\ LiCl$$

In practice, difficulties arise. The product was usually found to be a mixture of three and four-branch material. 1,2,4,5-Tetra (chlormethyl) benzene, under certain conditions[10], even produced a mixture of polymers some having as many as ten branches. Products containing molecules of functionality not greater than four could be obtained by adding the coupling agent slowly to a polystyryllithium solution in 50/50 benzene-tetrahydrofuran at 40–50 °C but even so the product had a mixture of functionalities (2, 3 and 4). Use of pure tetrahydrofuran and generally longer reaction times[11] other conditions remaining unchanged has been claimed to lead to complete coupling with both 1,2,4,5-tetra(chlormethyl)benzene and with hexa[*p*-chlormethylphenyl]benzene.

The presence of products having functionalities greater than theoretical is caused by competing metal-halogen exchange,

$$\sim\!\sim\!\sim^{-}Met.^{+} + RCl_x \rightarrow \sim\!\sim\!\sim Cl + RMet.Cl_{x-1}$$

since the product $RMet.(Cl)_{x-1}$ can now react with RCl_x to produce a new coupling agent of higher functionality. Some dimer will also be formed from the chlorine ended polymer. The effect of this reaction is minimized by keeping the active polymer chains in excess at all times. This problem seems to be particularly acute with benzylic halides but is not a major problem with other coupling agents particularly those based on silicon.

The presence in the product of molecules having less than the calculated number of branches is, however, a general feature of these reactions. A maximum of four branches rather than six could be coupled using bis(trichlorsilyl)ethane with polystyryllithium[12]. The cyclic trimer of phosphonitrilic chloride led to some material having a functionality of six[11, 12] but also to appreciable material of functionality two and four. Better results for $SiCl_4$ and CH_3SiCl_3 were claimed in the coupling of diene chains[13] but examination of the data suggests that the presence of some material of less than the calculated functionality cannot be excluded. Similar high degrees of coupling of diene chains were reported using dimethylphthallate or PCl_3[14]. From these reports it is nevertheless clear that the desired uniform product is not easily achieved. Much depends on reaction conditions apart from special difficulties associated for example with handling easily hydrolizable silicon halides, or possible side reactions[11] with phosphonitrilic chloride trimer.

Two basic problems exist

(i) Steric difficulties of coupling a number of bulky substituents if too many chlorine atoms are joined to a single atom e.g. $SiCl_4$. This difficulty is probably less severe if the chains are polydienes rather than polystyrene.

(ii) Once two chains have been coupled, the remaining chlorine atoms are essentially buried in a polymer coil. Subsequent reaction is diffusion controlled. Thus polystyryllithium for example reacts within the time of mixing with the first two chlorine atoms of 1,2-bis(methyldichlorsilyl)ethane even in benzene at room temperature[15]. Further reaction requires many days at 50 °C and becomes more difficult to force to completion the higher the molecular weight of the chains to be coupled. Local steric requirements are also still an important factor, for in forming the highest molecular weight ($\sim 10^6$) star polymers complete coupling could only be achieved by substituting a few diene units at the reactive end of the precursor chains.

While most of the attempts to form star polymers have involved chlorinated compounds as coupling agents, other compounds are available. The use of phthallate esters has already been mentioned. One particularly successful method involves tri(allyloxy)-1,3,5 triazine[16,17] or tetraphenyl-1,1,4,4 di(allyloxy)triazene-1,4 butane[18] for the preparation of 3 and 4-star polymers respectively. Titration of the active polystyrene chain produced in toluene/tetrahydrofuran at –70 °C with a toluene solution of the coupling agent until the red colour of the polystyrylpotassium disappeared led to near monodisperse products contaminated only with some precursor.

With all the methods of preparation, some fractionation of the product is unfortunately always necessary if only to remove excess precursor chains, which in most cases should be present to ensure complete use of the coupling agent and to help force reaction to completion. Precursor molecules are relatively easily removable but star polymers of lower functionality are not. Careful analysis of the product is always desirable. Measurements of number average molecular weights of precursor and product will give the correct degree of coupling. With higher molecular weight polymers the accuracy of osmotic measurements will be limited, so in this case weight averages can be used with little error if the polydispersity of the precursor is small. Sedimentation analysis under θ conditions will easily distinguish between material of functionality less than six[15], provided again that the chains are nearly monodisperse and that diffusional broadening is not important. The latter condition will exclude analysis of low molecular weight polymers. Gel permeation chromatography is often used in product analysis. No particular difficulty should arise in detection of the presence of linear chains (precursor and dimer). The effective volumes of star-branched polymers of equal arm length change little with functionalities of three or greater (see later sections). It seems likely therefore that the branched products of less than maximum degree of coupling will be difficult or impossible to detect. Experiments have been reported[19] for example where sedimentation analysis showed the presence of a number of products with varying degrees of coupling which were not obvious from the G.P.C. trace. For the same reasons a comparison of intrinsic viscosities of precursor and product will not be adequate to detect incomplete coupling.

Although it should be technically feasible to prepare star polymers of functionality somewhat greater than six by use of multifunctional silicon halides, this has so far not been attempted[1]. In this case, another technique has been used which involves

1 See, however Macromolecules *11*, 668 (1978)

divinylbenzene as coupling agent[20–22]. A few (usually 2–5) molecules of divinylbenzene are added to each precursor active chain by the normal anionic process. The modified chain ends now have both residual double bonds and organometallic groups which can react to link the chain ends. The coupling process is essentially self-limiting, since once several chains have been coupled, their reactive sites become "buried" in a central nodule. Internal reaction in the same structure then becomes more probable than reactions between groups on separate structures. The earliest polymers produced by this technique were made by using cumylpotassium initiated polystyrene in tetrahydrofuran. The products were rather polydisperse and contained appreciable amounts of the precursor molecules. Better results were obtained by working in benzene with sec-butyllithium initiated polystyrene[23]. It can be shown[24] that under these conditions with *p*-divinylbenzene, the reactivity of divinylbenzene with the organo-lithium compounds is ten times faster than that of the residual double bonds produced by coupling. The first rapid reaction consists therefore of addition of the divinylbenzene units to the chain end:

$$\mathrm{Bu}\{\mathrm{CH_2{-}CH}(\phi)\}_n\mathrm{CH_2\bar{C}H}(\phi)\mathrm{Li^+} + \mathrm{m}\ (para{-}\mathrm{dvb})$$

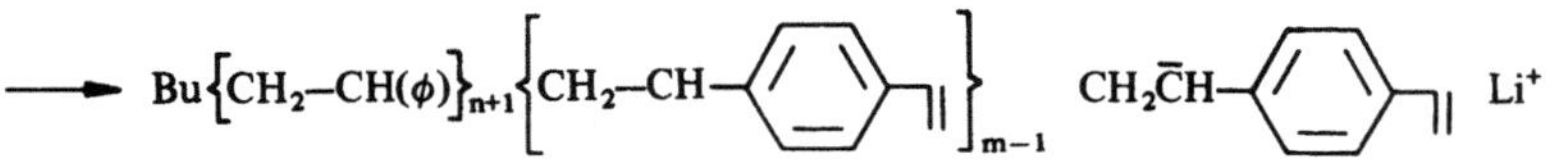

This process can be monitored spectrophotometrically[23]. It is followed by a slower cross-linking reaction leading to the spectrum characteristic of

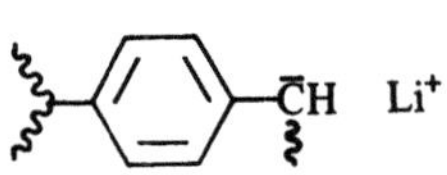 groups.

With *m*-divinylbenzene the reaction steps are not as well differentiated since the spectra of both species appear immediately on addition of the cross-linking agent.

The degree of coupling attained is a complex function of reaction variables[23,25]. Functionality increases as precursor molecular weight is decreased, as total polymer concentration is increased or as the amount of divinylbenzene is increased. The quantity of residual precursor molecules in the product is usually low (<3%) and although the samples produced by this method are probably not as monodisperse as indicated by sophisticated fractionation techniques[26] due to difficulties in separation of branched polymers of high molecular weight, $\bar{M}_w/\bar{M}_n$ is likely to be less than 1.3. It is in any case the only method available for the production of star macromolecules of high functionality and is useful for the preparation of macromolecules of functionality between six and twenty.

A variant on this technique has been used to prepare star polymers of extremely high molecular weight and functionality[27]. This involves direct addition of butyllithium to divinylbenzene in dilute solution. In a 2.5% solution for example, initial charges of up to six moles of divinylbenzene to one of butyllithium can be used without the occurrence of gelation. Soluble, densely cross-linked polydivinylbenzene particles of elevated molecular weight and functionalities as high as thousands can be obtained by this method. Since the nuclei still contain active organolithium groups, monomer at least can still diffuse into the nuclei to produce long chain branches. Heterogeneity is unfortunately larger than in the original technique involving addition of divinylbenzene to polystyryllithium. The two methods can, in fact, be considered as limiting cases of the "copolymerization" of styrene and divinylbenzene. A whole range of intermediate conditions can be used[25] where the ratio of divinylbenzene units to styrene units is allowed to increase from ~1% to 40% by decreasing the degree of polymerization of the precursor from ~200 to ~10. As the percentage of divinylbenzene is increased, a rapid increase in functionality occurs together with elevation of the molecular weight of the star polymer produced. In the original method only a very small percentage of divinylbenzene is used and the dependence of functionality on precursor molecular weight is relatively small. The fraction of material in the form of a highly crosslinked central nodule is only a few percent of the total material, but at high divinylbenzene ratios, 20–30% of the material can be in this form. The physical properties of the star polymer may be affected by this phenomenon.

The star macromolecules described so far have been homopolymers of styrene or the dienes. It will be obvious that with anionic polymerization techniques the active precursor molecules can be block-copolymers and that coupling by any of the methods described will yield radial (star) block-copolymers. Preparation of such polymers has been described in the literature[14, 28]. It is also possible to prepare star-copolymers having a mixture of branches. Star polystyrenes for instance, formed by the divinylbenzene technique can be used to initiate diene polymerization[25]. Since each coupled polystyrene chain leaves one organo-metallic group in the nucleus, polymerization of a diene in this way will lead to equal numbers of styrene and diene branches. Even using silicon-chlorine compounds, formation of mixed star polymers should be possible. Because of the large differences in reactivity between the first two chlorines of 1,2-bis(methyl dichlorsilyl)ethane and the rest, addition of half the stoichiometric amount of polystyryllithium followed by an equivalent amount of polyisoprenyl – (or polybutadienyl)-lithium should yield a reasonably homogeneous mixed star copolymer.

2 Calculations of the Effect of Branching Using Models

One of the most fundamental properties of a macromolecule is its size, normally measured as the mean square radius of gyration $\langle S^2 \rangle$. The conventional model for a polymer molecule consists of a random flight chain of n freely jointed segments each of effective length b which leads to the well known result $\langle S^2 \rangle = b^2 n/6$. The real chain unlike the model has restricted rotation about bonds and intersections are forbidden because of the volume of the chain units. It has been amply demonstrated that the differences between real and abstract chains can be discussed in terms of short and long range effects. Definition of an effective link length equivalent to several bonds circumvents the former effect. The long range (volume exclusion) effect can be nullified at a particular temperature by the use of solvents or their mixtures having segment-segment interactions more favoured than segment-solvent interactions (the so-called θ point). Under these conditions $\langle S^2 \rangle_\theta$ is found to be proportional to molecular weight as required by the model[29]. The effect of branching on $<S^2>$ is most simply discussed at the θ point. Coupling several random flight chains to a common point provides the required model. The effect on $<S^2>$ was evaluated[30] many years ago in terms of the ratio (g) of the mean square radius of gyration of the branched polymer to that of a linear polymer of the same molecular weight. For monodisperse chains all of equal length linked together[30] ("regular stars")

$$g = (3\,f - 2)/f^2 \tag{1}$$

where f is the functionality or number of side chains. The model would be appropriate for star molecules produced under ideal anionic polymerization conditions where the coupling is essentially complete. In some cases as mentioned earlier, these conditions may not apply. Burchard[31-33] has investigated the effects of

a) incomplete coupling
b) polydispersity of side chains and
c) finite size of the nucleus to which the chains are coupled.

The generalized results are complex so only a few extreme cases will be described, restricted as is Eq. (1) to polymers of reasonably high molecular weight and to a point nucleus.

1) Monodisperse side chains but the coupling is incomplete, i.e. only a fraction α of the actual number of reactive groups (f) has been utilized on the coupling agent.

$$g = \frac{1 + 3(f - 1)\alpha}{[1 + (f - 1)\alpha]^2} \tag{2}$$

2) For side chains having a "most probable" molecular weight distribution:

$$g_{z,w} = 4\,f/(f + 1)^2 \tag{3}$$

The comparison of $<S^2>_Z$ is made with a linear polymer of most probable distribution and the same *weight* average molecular weight. Equation (3) would be applicable to light scattering determinations. Due to the marked sharpening of molecular weight distribution on coupling chains of this type a number of g parameters could be defined. For example

$$g_{w,w} = 6\,f/(f + 1)(f + 2) \tag{4}$$

if both $<S^2>$ and M are weight averages. This expression holds for all average values of g (n, w, z) if the comparison of $<S^2>$ is made at the same type of average molecular weight. Typical curves showing g as a function of branching are given in Fig. 1.

If the nucleus is not of negligible size as will occur for some divinylbenzene coupled polymers and all the glycogen and amylopectin star molecules, then the expressions derived for g are complex. For a treatment of these particular problems the reader is referred to the original article[31].

Values of g can be determined by light-scattering measurements on the relevant linear and branched polymers. The function $P_{(\theta)}$ describing the angular dependence

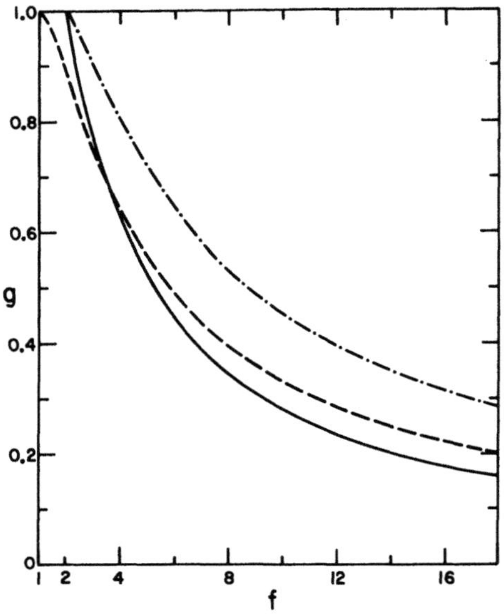

Fig. 1. Random flight calculations of g.
——, monodisperse side-chains [Eq. (1)].
– – –, $g_{z,w}$ for side-chains having a most probable distribution [Eq. (3)].
–·–·–, $g_{w,w}$ for side-chains having a most probable distribution [Eq. (4)]

of scattering gives directly, at small angles, the radius of gyration of all types of molecule;

$$P_{(\theta)} = 1 - h^2 \langle S^2 \rangle_Z / 3 + \ldots\ldots \quad (5)$$

where h is an optical constant involving the angular function $\sin^2\theta/2$. At higher angles, the scattering function is sensitive to molecular architecture. Figure 2 shows the calculated values of uP(u) where $u = h^2 \langle S^2 \rangle_Z{}^2$ for monodisperse polymers obeying random-flight statistics[34]. A maximum occurs in this function with branched polymers having four or more arms. Calculations have also been made for polydisperse chains and for other types of branching[32]. In principle more detailed information is available on segment distribution from the overall angular function, since at higher u values, $P_{(u)}$ becomes sensitive to smaller intramolecular distances. High u values are however not available using light-scattering techniques which are restricted to $u^{1/2} < \sim 5$ even for polymers having molecular weights of several millions. Neutron scattering measurements would be necessary for delineation of greater ranges of u.

Virtually all the experimental data available deals with regular monodisperse star molecules prepared by anionic polymerization, where either by fractionation or by experimental design, coupling is essentially complete. Consideration will be henceforward normally restricted to this special case. The mean segment density of star-branched polymers is greater than for linear polymers of the same molecular weight as indicated by the g factor. Even compared at the same radius of gyration the higher density remains. The average radial segment distribution is very similar to that of a linear polymer[35] but some differences can be calculated in the instantaneous distribution which becomes more symmetrical for regular star polymers[36].

---

2 Note the parameter u is defined in various ways in the literature.

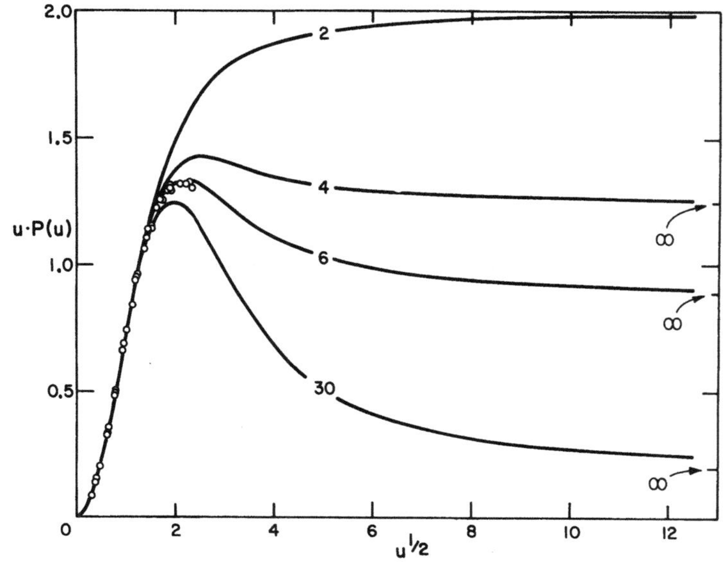

Fig. 2. Angular scattering functions for regular star-branched polymers with f = 2, 4, 6, 30. Experimental points, six-branched star polyisoprene (Ref. [57])

The distribution about the mean value of radius of gyration also sharpens. It is therefore unlikely that the observed effects on branching can be accounted for simply by the g factor.

Early experimental evaluations of the solution properties of branched polymers[37] did in fact assume that the Φ′ and P′ constants of the standard equations for intrinsic viscosity and frictional coefficient of flexible linear polymers;

$$[F]_\theta = P' \langle S^2\rangle_\theta^{1/2} \tag{6}$$

$$[\eta]_\theta = \Phi'(\langle S^2\rangle_\theta/M)^{3/2} M^{1/2} = K_\theta M^{1/2} \tag{7}$$

would remain unchanged, differences for branched polymers being only caused by the reduced values of $\langle S^2\rangle$. If we define

$$g' = [\eta]_\theta^{star}/[\eta]_\theta^{linear} \text{ and } h = [F]_\theta^{star}/[F]_\theta^{linear} \tag{8}$$

comparisons being taken at the same molecular weight, then g′ would be expected to be equal to $g^{3/2}$ and h to $g^{1/2}$.

Values of Φ′ and P′ can be calculated using the "bead-spring" model of polymer molecules in which the chain is divided into sub-units obeying random flight statistics and acting as entropy springs, joined by beads undergoing frictional interaction with the solvent. Two extreme cases can be envisaged, the first evaluated by Rouse involves no disturbance to solvent flow ("free-draining")[38] the second discussed by Zimm assumes extensive hydrodynamic interaction[39] ("non-free-draining"). The latter case has been shown[29] to apply to dilute solution flow properties, for it leads directly to Eqs. (6) and (7). The effect of various types of branching on hydrodynamic behaviour of polymer solutions has been investigated both in the free-draining[40, 41] and non-free-draining[40] limits.

For regular star-branched polymers under the non-free-draining conditions applicable for Eqs. (6) and (7), no exact closed expression could be derived for g′ but approximately[40]:

$$g' \approx \left(\frac{2}{f}\right)^{3/2} [0.390(f-1) + 0.196]/0.586. \qquad (9)$$

Exact numerical solutions are available for certain values of f[42]. An exact solution is possible for h[35];

$$h = f^{1/2} [2 - f + 2^{1/2}(f-1)]^{-1} \qquad (10)$$

Values of g, g′ and h for regular star-branched polymers are given in Table 1. It will be observed that g′ and h are always closer to unity than predicted from the simple argument presented above. The higher segment density in such polymers affects the flow properties, both Φ' and P′ increasing with branching. Φ' is particularly sensitive, increasing quite rapidly from the normal value of 3.9×10^{22} (dl. g^{-1}).

The model also predicts the energy stored during steady state deformation in a shear gradient. The steady state compliance J_e^0 is expected to decrease quite rapidly with branching. Values of $g_2 = J_e^0(\text{star})/J_e^0(\text{linear})$ at constant M, are given in Table 1. The relaxation spectrum of regular star macromolecules is predicted to move to shorter times compared with linear molecules and the odd-numbered modes to become (f – 1) fold degenerate. The longest relaxation time τ_1, which makes the largest contribution to $[\eta]$ and J_e^0 is predicted to decrease by a factor of $(2/f)^{3/2}$. From the relaxation spectra the form of the frequency dependence of the shear storage (G′) and shear loss (G″) moduli can be calculated. At low frequencies both [G′] and [G″], the intrinsic moduli[3], are predicted to decrease on branching, the latter by the factor g′ and the former by a larger factor. A more detailed discussion of the application of the bead-spring model to viscoelastic properties of solutions of linear and branched polymers is available in an earlier issue of this series[43].

3 $[G'] = \text{limit}_{c \to 0} G'/c$, $[G''] = \text{limit}_{c \to 0}(G'' - \omega\eta_{\text{solvent}})/c$.

Table 1. Calculated values of g, g′, h and g_2 for regular star-branched polymers (non-free draining limit)

f	g	g′ [Eq. (9)]	$g^{1/2}$	$g^{3/2}$	h	g_2
3	0.778	0.907	0.882	0.686	0.947	0.684
4	0.625	0.824	0.790	0.494	0.892	0.514
6	0.444	0.705	0.667	0.296	0.798	0.342
8	0.344	0.624	0.586	0.202	0.725	0.256

In concentrated solutions (or molten bulk polymer) the free-draining version of the model is more appropriate[44], although difficulties arise in its use because it strictly only applies in the absence of inter-molecular interactions. For regular stars, g′ is simply equal to g itself under these conditions[40, 41]. The concentrated solution viscosity will obey an equation of the form[44]

$$\eta = K'\langle S^2\rangle \tag{11}$$

where K′ is independent of molecular weight but depends on density (or polymer concentration) and the frictional coefficient of an individual bead. A somewhat larger decrease in viscosity is therefore predicted for a given degree of branching than is observed for intrinsic viscosities (Table 1). A decrease in J_e^0, the steady state compliance, is also predicted on star-branching[41] in concentrated solution. Equation (11) is only obeyed however at moderately low molecular weights. Above a critical molecular weight which depends on the polymer concentration, entanglements between polymer chains become important, leading to two extra energy loss terms which increase the viscosity. Each moving molecule will tend to pull others along with it and in addition each molecule must slip through the entanglement points[44, 45]. In this molecular weight range the viscosity becomes more dependent on molecular size[46] i.e.

$$\eta = K\langle S^2\rangle^{7/2} \tag{12}$$

where K depends on the number of segments between entanglement points and includes a factor to account for the strength of entanglement coupling between molecules. Both these factors should depend on the nature of the polymer chain but should be constant for a given polymer species. Calculations of the effect of star-branching on such a flow mechanism[47] lead to the conclusion that the basic formula will be unchanged provided that the entanglement spacing is not changed by branching. The ratio $\eta_{star}/\eta_{linear}$, equal to g at lower molecular weights, will simply become $g^{7/2}$ at higher molecular weights which will produce a much more drastic drop in viscosity than in dilute solution. The breakpoint (M_b) between the two regions should moreover move to higher molecular weights as the number of branches increases i.e. $M_b \sim g^{-1}$.

3 Dilute Solution Properties

Reported data on g and g′ for regular star polystyrenes and polyisoprenes are collected in Table 2. In examination of the results it should be noted that substantial variation exists in reference parameters for linear polymers. For example the K_θ used for normal polystyrene ranges from 7.45×10^{-4} to 8.4×10^{-4}. Much of this variation probably reflects differences in calibration standards in the determination of light scattering molecular weights. In the cases where no data are reported on linear polymers it must be assumed that the value chosen reflects the laboratories previous

Table 2. Experimentally determined values of g, g′ of regular star-branched polystyrene

Ref.	g(θ)	g′(θ)	g′(benzene)	g′(toluene)
f = 3, calculated g = 0.778				
8)	–	0.94(1)	–	0.90(1)
48)	–	0.85(1)	$0.87_4(1)$	0.88(1)
17)	–	$0.83_5(5)$	0.81(5)	–
f = 4, calculated g = 0.625, g′ = 0.814				
8)	–	0.82(1)	–	0.84(1)
50) a	0.58(1)	0.75(1)	–	–
15, 49)	0.63(5)	0.76(9)	–	$0.72_5(9)$
51)	–	0.74(6)	0.70(6)	–
52)	0.64(1)	0.77(1)	–	0.68(1)
f = 6, calculated g = 0.444				
52) b	$0.42_5(2)$	0.57(2)	–	0.46(2)
49)	0.46(3)	0.63(9)	–	0.57(9)
Multibranch[53]				
f = 8.7	0.31	0.45	0.44	–
10.7	0.27_5	0.39	0.40	–
12.3	0.26	0.36	0.36	–
15.3	0.23	0.30	0.30	–
Values for polyisoprene[54]				
f = 4	0.65(3)	0.77(7)	–	0.73(7)
6	0.46(3)	0.63(7)	–	0.59(7)

a Measured f = 4.75, g calculated = 0.54.
b Calculated using $\bar{M}_n$ values of the stars which suggest f ~ 7. Use of $\bar{M}_w$ values would increase the values of g, g′.
Numbers in parentheses indicate number of polymers characterized.

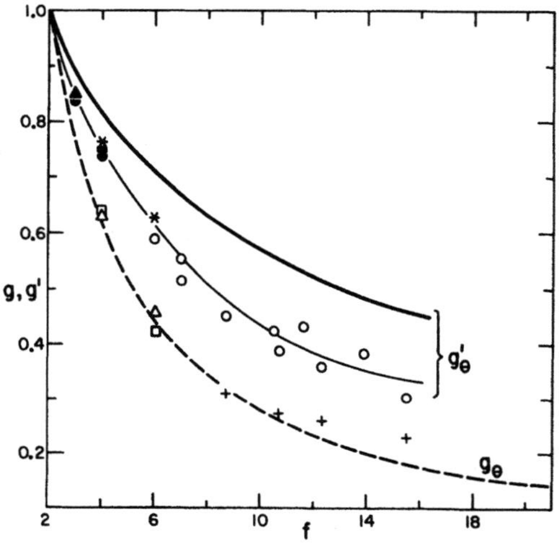

Fig. 3. Experimentally determined values of g_θ and g'_θ for star-branched polystyrene. g_θ, □ Ref.[52], △ Ref.[15, 49], + Ref.[53], ----Eq. (1). g'_θ, ▲ Ref.[48], ● Ref.[17, 51], ∗ Ref.[15, 49], ○ Ref.[53], heavy line is theoretical

experience with linear polymers. The inclusion of some data on linear polymers enhances the confidence in the reported g and g′ values.

With these reservations in mind, it seems that at least up to eight or nine branches, Eq. (1) predicts, within experimental accuracy, the decrease in radius of gyration on branching (Fig. 3). Above this point some divergence occurs between theory and experiment. In addition, the temperature at which the second virial coefficient becomes zero decreases below the normal θ temperature to an increasing extent even for polymers of molecular weight greater than 10^6. Deviations from θ occur even for four and six-branched star polymers of low molecular weight but their radii of gyration are not measurable by light-scattering techniques. The values of g_θ reported for these polymers refer to those of high molecular weight where ambiguity does not arise. No such solution is possible for more highly branched polymers, so all values of g_θ are standardized to the measured value of $\langle S^2 \rangle$ at the temperature where $A_2 = 0$. Benoit and co-workers[55)] have suggested that deviations from Eq. (1) for more highly branched polymers are due to the chains being still slightly expanded under conditions where the second virial coefficient is zero. Thus for polymers of abnormally high segment density the temperature at which the chain expansion coefficient is unity (θ_α) and at which $A_2 = 0$ (θ_{A_2}) are different from the normal θ temperature i.e. $\theta_\alpha < \theta_{A_2} < \theta$. An alternative explanation could be presented in terms of changes in the effective segment length. A Monte-Carlo lattice calculation of radii of gyration of star-branched polymers[56)] incorporating both physically excluded volume and an attractive energy for nearest-neighbour segments predicts little change in the θ-point on branching (defined as that where random flight statistics are obeyed) but a significant increase in the effective segment length b of individual arms. g_θ values calculated are about 0.05 higher than predicted by Eq. (1). This seems to be too large a correction at least for the polymers with lower numbers of arms.

In addition to conventional light-scattering measurements at low angles to determine mean square radii of gyration, attempts have been made to check the form of the angular scattering function over a wider range of u values. Only polymers of very high molecular weight are suitable for this type of measurement since $u = h^2 \langle S^2 \rangle_z$. To further increase $\langle S^2 \rangle_z$, measurements were made in good solvents rather than the θ-solvents required for strict application of the theory. Nevertheless the results on a near-monodisperse six-branched star polyisoprene ($\bar{M}_w = 1.7 \times 10^6$) are in good agreement with theory up to the maximum in uP(u) as shown in Fig. 2[57)]. A somewhat wider range in u could be obtained[32)] with a star-branched polystyrene of even higher molecular weight ($\bar{M}_w = 16.7 \times 10^6$) in toluene. In this case the experimental data fitted the theoretical curves for seven to nine branches although other methods of characterization suggested a much larger number of branches. Some polydispersity in the sample, prepared by the divinylbenzene technique, is the probable cause of the discrepancy. Unfortunately the maximum value of uP(u) is quite sensitive to this factor.

The predictions of the Zimm-Kilb theory for g′ [Eq. (9)] are in relatively poor agreement with experiment even for four and six-branched regular star polymers, the most favourable case for the preparation of well-defined polymers. The discrepancy becomes larger with higher branch numbers although some scatter is apparent in the

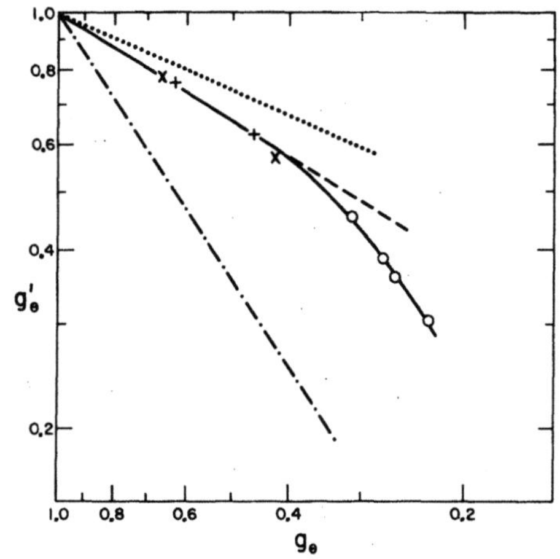

Fig. 4. Relationship between g_θ and g'_θ. ...... Zimm-Kilb Theory —·—·— $g' = g^{3/2}$. + Ref. [15, 49]. x Ref. [52]. o Ref. [53]

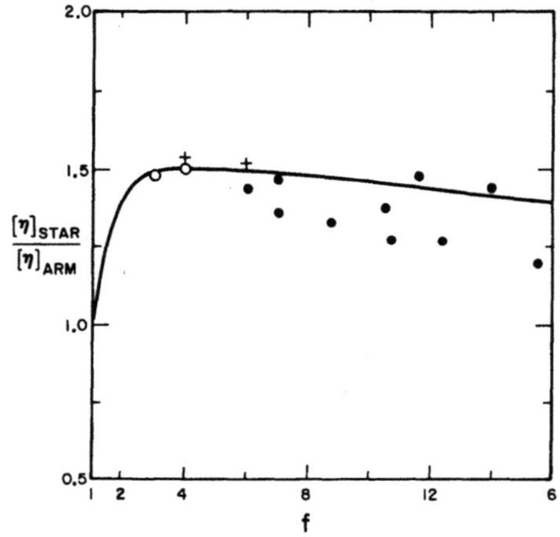

Fig. 5. [η] of star-branched polystyrene of a common arm length. Effect of increase in the number of arms. —— calculated for $g' = g^{0.61}$ o Ref. [17, 51]. + Ref. [15, 49]. ● Ref. [53]

results (Fig. 3). It should be noted that for f > 8, the theory predicts Φ′ values greater than for a uniform density sphere (13.5×10^{22}) which does not seem reasonable. In practice the Φ′ values appear to increase quite rapidly with branching, but to reach a plateau value near three times Φ'_{linear} i.e. slightly below the rigid sphere value for large values of f. The theoretical values of g′ can be expressed in terms of g by the simple formula $g' = g^{0.44}$ as pointed out by Berry[50]. Figure 4 shows the experimentally observed relationship between the two parameters under θ conditions plotted logarithmically. The slope is close to 0.60 for the lowest members of the series but appears to increase at higher branch numbers suggesting that a simple power law will not hold under all conditions.

It should be noted that if $g' = g^{0.5}$, then intrinsic viscosity is simply a function of mean square radius of gyration alone. The theoretical treatment suggests that intrinsic viscosities of star-branched polymers will be slightly higher than linear polymers of the same mean square radius of gyration. The experimental results, however, indicate that the branched polymers have lower intrinsic viscosities when compared on this basis.

Table 3. θ_{A_2} temperatures for regular star-branched polystyrenes in cyclohexane[49, 53]

a) 4-Star	10^{-4} M_w	5.02	9.33	15.2		
	θ_{A_2} (°C)	29.2	31.8	32.1		
b) 6-Star	10^{-4} M_w	6.6	10.9	22.5	31.7	59.1
	θ_{A_2} (°C)	25.3	27.8	30.2	32.0	33.4
c) 12–13 Star	10^{-4} M_w	9.5	22.3	23.5		
	θ_{A_2} (°C)	18	25	30		

θ_{A_2} temperatures for regular star-branched polyisoprenes in different solvents

		Solvent	ψ_1	$(1/3 - \chi_2)$	$(\theta - \theta_{A_2})$
a)	$M_w = 45{,}000$[54] f = 6	Methylisobutylketone	0.41	0.027	2.8°
		Dioxane	0.35	–	0°
		Methyl-n-propylketone	0.18	0.027	6.4°
b)	$M_w = 342{,}000$[59] $\bar{f} = 22$	Methylisobutylketone	–	–	1.5°
		Methyl-n-propylketone	–	–	9.5°

The addition of extra branches of a common length has very little influence on either radius of gyration or intrinsic viscosity[53] once a few branches are present. For example with a common arm molecular weight of 190,000 the unperturbed root mean square radius of gyration of star polystyrenes with an average of 10.7, 12.3 and 15.5 branches is 267, 278 and 293 Å respectively; $[\eta]_\theta$ decreases slightly in the same series (0.422, 0.417, 0.395 dl. g^{-1}). This type of result follows directly from chain statistics of star polymers. $\langle S^2\rangle_{star}/\langle S^2\rangle_{arm}$ is given by $(3f - 2)/f$, which only increases from 2.33 for a three branch polymer to 2.8 for a ten branch one, with a limit of 3.0 for an infinite number of branches. Similar results are predicted for intrinsic viscosity. From the definition of g′ [Eq. (8)], $[\eta]_{star}/[\eta]_{arm} = g'f^{0.5}$. If $g' \sim g^{0.5-0.6}$ then the viscosity ratio will be in the range 1.4 to 1.5 for polymers with more than three branches. The actual values predicted with $g' = g^{0.6}$ are shown by the line in Fig. 5 together with experimental results for a number of polymers.

The deviations of θ_{A_2} from the value observed for linear polymers (θ) provide an interesting example of the failure of standard theories of dilute polymer solutions. $\theta - \theta_{A_2}$ is larger the greater the number of branches and the shorter is the arm length (Table 3) i.e. the difference appears to correlate with increased segment density. Calculations of A_2 using the smoothed segment density model of polymer chains for example evaluate the change in free energy involved in overlapping two chains[29]. Terms involving higher powers than two in segment concentration are neglected in the expansion of the free energy function. Orofino and Flory[58] and Benoit and co-workers[55] have modified this approach to include cubed terms i.e. the local excess chemical potential becomes

$$(\mu_1 - \mu_1^0)_{excess} = -RT\left[(\psi_1 - \kappa_1)v_2^2 + \left(\frac{1}{3} - \chi_2\right)v_2^3 \ldots\right] \quad (13)$$

Under conditions where the heat and entropy parameters κ_1 and ψ_1 cancel (the θ

point) $\left(\frac{1}{3} - \chi_2\right)$ may still be finite and important at high segment concentrations[4]. A_2 will still be positive at θ under these conditions.

When modified in this way the theory predicts

$$(\theta - \theta_{A_2})/\theta_{A_2} \sim \left(\frac{1}{3} - \chi_2\right) H/A^3\alpha^3 g^{3/2} M^{1/2} \psi_1 \tag{14}$$

where θ is the normal Flory temperature, H is a function which is unity (Orofino-Flory) or slightly above unity (Benoit), A^3 is the usual parameter $(\langle S^2\rangle_\theta/M)^{3/2}$ for linear chains and α the expansion coefficient at θ_{A_2}. As can be seen, the θ_{A_2} depression is correctly predicted to increase with branching via the $g^{3/2}$ term and to decrease at higher molecular weight. Moreover, the magnitude of the effect should vary between different θ solvents, for ψ_1 is often different in such cases.

The best fit to data on branched polymers can be made by choice of values of $(1/3 - \chi_2)$. Figure 6 shows such a fit to $(1/3 - \chi_2) = 0.045$ for a series of four and six-branched regular star polymers. Semi-quantitative agreement is achieved although the θ depression disappears more rapidly as molecular weight is increased than is predicted. The effect of solvent on deviations of θ_{A_2} from θ has been investigated for

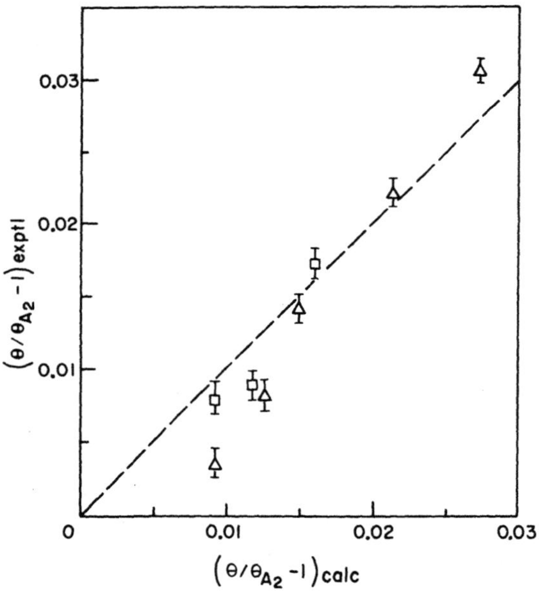

Fig. 6. Depression of θ_{A_2} for regular four and six-branched star polystyrenes in cyclohexane. ⌶ 4-star. ⌶ 6-star (Ref. 49)). Calculated values from Eq. (14). [Reproduced from Macromolecules 7, 443 (1974).]

---

4 Strictly the extra term should, as the first is, be divided into heat and entropy terms and hence be temperature dependent. As a first approach temperature variation of $\left(\frac{1}{3} - \chi_2\right)$ is neglected.

star polyisoprenes[54, 59]. $\theta - \theta_{A_2}$ is greater for methyl-n-propylketone than for methyl-isobutylketone which correlates with their different ψ_1 values. $(1/3 - \chi_2)$ is essentially identical in the two solvents. In dioxane the θ depression is zero apparently because the $(1/3 - \chi_2)$ term is completely negligible. It seems, therefore, that dilute solution properties of star-branched polymers can be at least approximated in terms of three parameters ψ_1, θ and χ_2. The smoothed density model is known to predict the variation of α and A_2 rather poorly for linear polymers so that quantitative agreement for branched polymers is unlikely.

In the above discussion it is implicitly assumed that ψ_1 (as well as θ) has the same value as for linear polymers. ψ_1 is normally determined from the variation of A_2 with $(1 - \theta/T)$ in a very small temperature range arount θ where A_2 is simply proportional to $\psi_1(1 - \theta/T)$. For some of the polymers studied $A_2 \neq 0$ at θ which raises some doubts about the procedure under these conditions. The Orofino-Flory modification, which appears to be the more exact[49], suggests in this case that θ should be simply replaced by θ_{A_2} if the two are different. Figure 7 shows the variation of A_2 around θ_{A_2} for linear, four-branched and six-branched polystyrenes where this procedure has been applied[5]. The accuracy is limited because of the small usable temperature range, but no evidence exists for change in slope and hence variations in ψ_1 greater than 10%. Berry[50] has reported similar results for a four-branched and a six-branched polymer in decalin.

Not all the reports in the literature, however, suggest a normal entropy parameter for star-branched polymers. $\psi_1 = 0.25$ was reported for a three-branched poly-

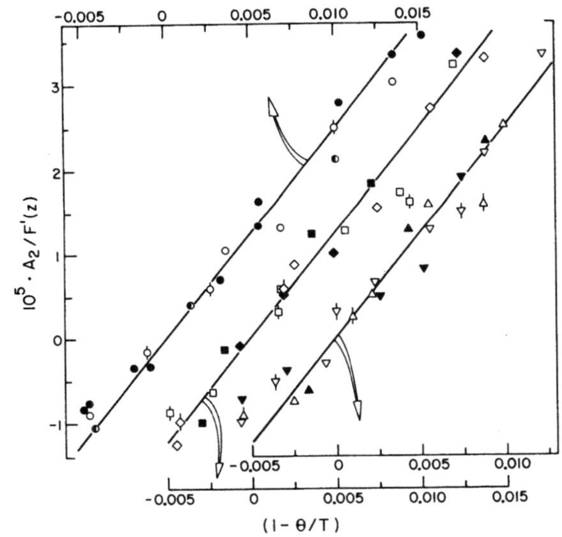

Fig. 7. A_2 versus $(1 - \theta/T)$ for linear and regular four and six-branched star polystyrenes in cyclohexane. Circles, linear. Squares 4-star. Triangles 6-star. The F'(z) correction is described in Ref.[66] It increases the linear range. [Reproduced from Macromolecules 7, 443 (1974).]

5 θ and θ_{A_2} are identical for many of the polymers shown, but a longer linear range is available for the polymers of lower molecular weight where $\theta \neq \theta_{A_2}$.

styrene using the same technique[48]. This is slightly below the normal value. ψ_1 values determined indirectly by the application of the modified smoothed density model are, however, considerably lower (~0.03[55]). This model is known to give low values even for linear polymers but the error is approximately a factor of two. It can be made to yield correct values by changing the constant involved (C_m in Flory's notation) to force a fit with exact perturbation theory valid at low chain expansions[60]. Use of the modified smoothed density model on those branched polymers of high molecular weight ($\theta_{A_2} = 34.5°$) shown in Fig. 7 would also lead to abnormally low ψ_1 values (<0.1), in contrast to the normal values found directly from A_2 near θ. The low ψ_1 values are determined from the equation,

$$\alpha^5 - \alpha^3 = 2\,C_m\psi_1\left(1 - \frac{\theta}{T}\right)M^{1/2}g^{-3/2} + 2\,C'_m/3\,\alpha^3 g^3 \tag{15}$$

When α is close to unity this equation will predict strongly enhanced values of the temperature coefficient of α^2 for branched polymers because of the presence of the term $g^{3/2}$, in poor agreement with perturbation theory[61] which predicts quite small effects. For a six-branched star polymer for example Eq. (15) predicts a 300% increase while perturbation theory gives only 13%. It seems that ψ_1 values obtained in this manner will be seriously underestimated.

Some data are available on the behaviour of regular star macromolecules in good solvents (Table 2) but most of it refers to g′ values. Some scatter is evident in the experimental results. A few reports exist where g′ is higher in a good solvent than under θ conditions but for the majority of well characterized polymers a small decrease is observed.

The measurements on g′ for regular star polymers in good solvents can be compared with those reported for star-polycondensates having a most probable distribution in branch length[1]. In the latter case although a quite large decrease in $[\eta]$ was observed for four and eight branch polymers when compared at the same number-average molecular weight, this was shown to be largely the result of the changing molecular weight distribution with branch number. When compared at the same viscosity average molecular weight to correct for this factor, a very small decrease in $[\eta]$ was observed for the four-branch samples and only a ~15% decrease even for the eight-branch ones. While the polymers were not ideal for such a comparison, (the number of monomer units per branch was as low as four and never greater than forty) at least some of the smaller effect of branching in this case could arise because the decrease in average square radius for a star-branched polymer of a given number of total segments but randomly distributed among the branches is much lower than for regular star polymers. The g factors are in fact given by Eq. (4) in the former case and Eq. (1) in the latter[62]. The relative magnitudes are illustrated in Fig. 1. g′ would be expected to be even closer to unity than g itself. No evaluation of g′ for polymers with randomly distributed arm lengths appears to have been made but values of h have been calculated[62]. h does indeed decrease less rapidly with branching than for the case of equal arm lengths. To obtain agreement between theory and experiment on g′ however the calculated g′ difference between random and uniform branched polymers would have to be even larger then is found for h.

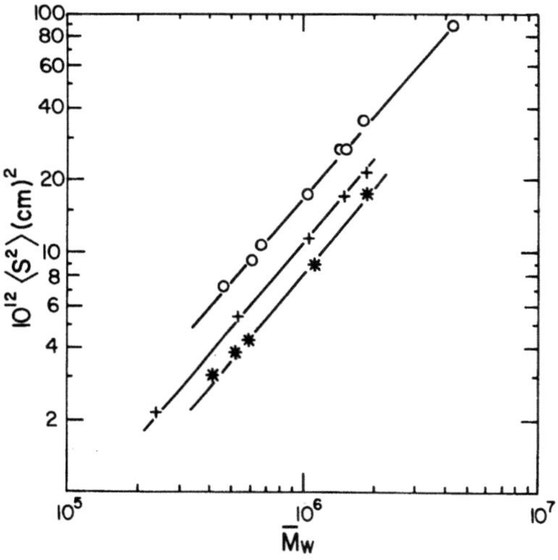

Fig. 8. Mean square radius of gyration of polystyrenes in toluene. ○, linear. +, 4-star, *, 6-star. Data from Refs.[52, 65, 66]

Dynamic viscoelastic measurements can be made at very low concentrations and extrapolated to infinite dilution to obtain intrinsic storage and loss moduli, [G′] and [G″] of the isolated polymer chain. Measurements have been made on a nine-branch star polystyrene and a four-branch star polybutadiene in several solvents including two θ solvents in the former case and over a frequency range of about 100 to 6000 Hz[63, 64]. At low frequency [G″] will be in error since $[\eta]$ $(= [G'']/\omega\eta_s)_{\omega \to 0}$ is lower than predicted. Essentially the experimental results are normalized to the observed value to avoid this difficulty. The observed form of the frequency dependence is adequately represented by the Zimm-Kilb theory[40] and its modifications for use in good solvents using a single parameter h* which is a function of the hydrodynamic interaction between segments. This suggests that the form of the relaxation spectrum is close to that suggested by theory. Significant differences are expected and found between linear and branched polymers of the same molecular weight. [G′] (and the steady state compliance J_e^0) is very much smaller for the branched polymers in the low frequency region which can be correlated with the large decrease in τ_1 expected on branching.

Less data are available on thermodynamic properties in good solvents. Figure 8 shows the variation of $\langle S^2 \rangle$ with molecular weight for linear and regular star-branched polystyrene in toluene[52, 65, 66]. Mean g values of 0.65 and 0.48 can be derived for four and six-branched polymers respectively, which are not greatly different from the values obtained under θ conditions (Table 2). The values reported for multibranch polymers are however significantly higher in benzene than in cyclohexane[53]. No exact theory is available for expansion coefficients (α) in good solvents but Berry[50] has presented an approximate formula which suggests about a 5% increase in α^2 for a six-branched star polymer compared to a linear one. This is well within the experimental error in the determination of $g_{\text{good solvent}}/g_\theta$ required to measure the α^2 ratio.

Several approximate theories exist to describe second virial coefficients under conditions far removed from the θ point[67]. All predict A_2 will be of the form:

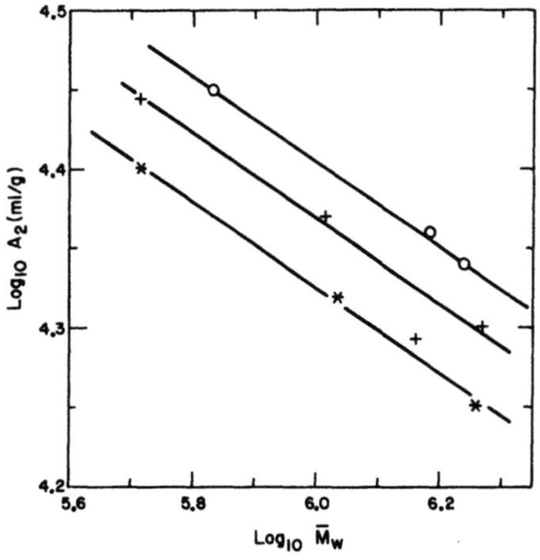

Fig. 9. A_2 for polystyrenes in toluene at 35°. Symbols as in Fig. 8, Ref. [65]

$$A_2M^2 = 4\,\pi^{3/2}N_a <S^2>^{3/2}\Psi \tag{16}$$

where Ψ, zero at the θ point, tends to a limiting value in very good solvents. This limit differs between the various theories. Under these conditions the polymer coil can be described as a non-interpenetrating sphere of radius proportional to $<S^2>^{1/2}$. If radius of gyration were the only determining factor, then A_2 at a given molecular weight should be decreased on branching by the factor $g^{3/2}$ i.e. 0.49 and 0.30 for four and six-branched polymers respectively. Figure 9 shows that A_2 in fact, in good solvents, is decreased by a much smaller factor, i.e. 0.9 and 0.8 in the two cases. This corresponds to Ψ values of 0.5 and 0.7 for four and six-branched polymers. Berry has reported similar results[50] ($\Psi_{linear} = 0.3$, $\Psi_4 = 0.56$, $\Psi_6 = 0.8$). It is evident that due to the increased segment density in branched polymers the equivalent sphere has increased in radius i.e. the effect of branching in decreasing radius of gyration has been offset by greater impenetrability.

One dilute solution property of practical interest is exclusion chromatography, widely used for characterization of molecular weight and its distribution. It is generally assumed that separation is governed by the size of the polymer molecule, the smaller molecules eluting later. Branched polymers are in fact retained on the column longer than their linear counterparts of the same molecular weight. Benoit and co-workers have suggested empirically that the elution volume is a unique function of the hydrodynamic volume as measured by the product $[\eta]M$[68,69]. Confirmation of the validity of this relationship was obtained for linear, star-branched and comb-branched polystyrene as well as for a number of other linear polymers. Casassa[70] has emphasized that branched polymers provide the best test of the universality of such relationships. Under the experimental conditions normally used for linear polymers, $[\eta]M$ will parallel alternative measures of molecular volume such as radius of gyration or end-to-end distance which have also been suggested as the definitive parameters. This follows because $[\eta]$ is expected to conform to the classical relationship $[\eta] = \Phi'<S^2>^{3/2}/M$ with a constant Φ' appropriate to the very good solvent normally used for elution. The chain expansion under most experimental conditions will also be fairly constant, so that even unperturbed dimensions could be used. We have

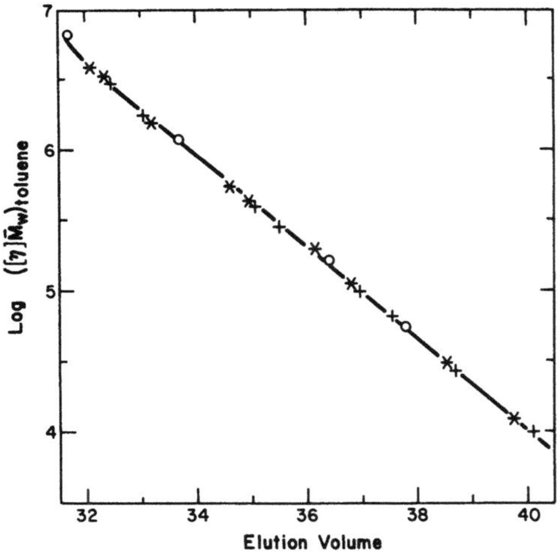

Fig. 10. G.P.C. experiments on linear and branched polystyrene in toluene at 35°. Symbols as in Fig. 8, Ref.[71]

seen earlier that Φ' does in fact change on branching so that discrepancies will be produced if the wrong measure of molecular volume is used. The results of Benoit and co-workers[68, 69] suggest $[\eta]$M is indeed the correct parameter to correlate with elution volume. Although no radii of gyration of the polymers are reported, it seems unlikely that with a wide variety of star and comb polymers $\langle S^2 \rangle$ would be a constant fraction of $[\eta]$M in all cases as unfortunately happened in the case of a few samples reported by Berry[50]. Figure 10 shows unpublished G.P.C. data[71] on well characterized four and six-branched star polystyrenes[49] in toluene for which the $[\eta]$M vs. $\langle S^2 \rangle$ relationship is definitely different for the three types of polymers and where the $[\eta]$M correlation works well. Use of $\langle S^2 \rangle$ (toluene) instead of $[\eta]$M as the correlation parameter would have resulted in a 0.25 log unit displacement between linear and six star polymers and an even greater difference if $\langle S^2 \rangle_\theta$ had been used. Deviations from the $[\eta]$M law have been observed for some comb-branched polymers[72, 73] but not in others[51, 69] so the position with regard to this type of polymer is unclear.

The use of hydrodynamic volume need not imply that flow properties are the important factor in exclusion chromatography. Casassa[70] has calculated the distribution of flexible macromolecules (linear and star-branched) between the bulk phase and small cavities assuming this is determined by the loss of conformational entropy of the molecules in the cavity. The distribution constant K_f was not a unique function of the total number of segments nor of the mean square radius of gyration but correlated best with the function $(Nb^2/6)^{1/2} g^{1/6} = \langle S^2 \rangle^{1/2}_{\text{lin.eqvt.}} g^{1/6}$. This is simply $([\eta]M/\Phi'_{\text{linear}})^{1/3}$ assuming the normal Fox-Flory viscosity relationship and $g' \approx g^{1/2}$. The latter relation is not exact for star-branched polymers as noted earlier (nor is the $g^{1/6}$ dependence) but the calculations do give a semi-quantitative explanation for the success of the $[\eta]$M law, for molecules with the same K_f should elute at the same point.

4 Concentrated Solution and Melt Flow Properties

Figure 11 shows zero shear viscosity data for regular linear, four and six-branched polystyrenes in diethylbenzene at a concentration of 0.255 g/ml[74]. The transition between a low molecular weight region where η approaches proportionality with M and a higher molecular weight zone with η approximately proportional to $M^{3.4}$ is seen. A decrease in viscosity on branching occurs as predicted by the Bueche theory[47] and the transition to a higher power dependence on molecular weight is delayed for the branched polymers. At the highest molecular weights, η for a 4-branch polymer is about 11% of its linear equivalent, dropping to about 3% for the six-branched material. These values are noticeably lower than expected from the $g^{3.4}$ relationship which predicts decreases to 20% and 6% respectively. It follows that the data are not well fitted to a single line by changing the abscissa to log (gM) as theory requires i.e. viscosities are equal at equal values of $\langle S^2 \rangle$. A more successful superposition can be attained by plotting $\log(\eta/[\eta]_\theta)$ against $\log[\eta]_\theta$ [74] a formula suggested by an extension to branched polymers of a corresponding states treatment of viscosities of linear polymers. An equivalent superposition could be attained by plotting log η against $\log(g^{1.22}M)$ since $[\eta]_\theta{}^2$ is simply $(g')^2K^2_{lin.}M$ and $g' = g^{0.61}$ for these polymers. This is, of course, the same function required to produce superposition of the dilute solution viscosity data for the same polymers. Both types of viscosity are thus more sensitive to $\langle S^2 \rangle$ than either theory suggests and surprisingly to about equal extents.

Similar results are obtained for linear, four and six-branched polyisoprenes at a concentration of 0.145 g/ml. In this case, however, at higher concentrations more serious deviations occur from theory. The higher molecular weight samples can have zero shear viscosities higher than linear polymers of the same molecular weight[75]. Such behavior was first noted in a study of melt viscosity of regular star-branched polybutadienes[76]. Viscosities of the order of one hundred times that of a linear equivalent could be observed, but the effect decreased rapidly on dilution with solvents[77] i.e. the viscosities of branched polymers were more sensitive to concentration than those of linear polymers. Star-branched polyisoprenes show viscosity enhance-

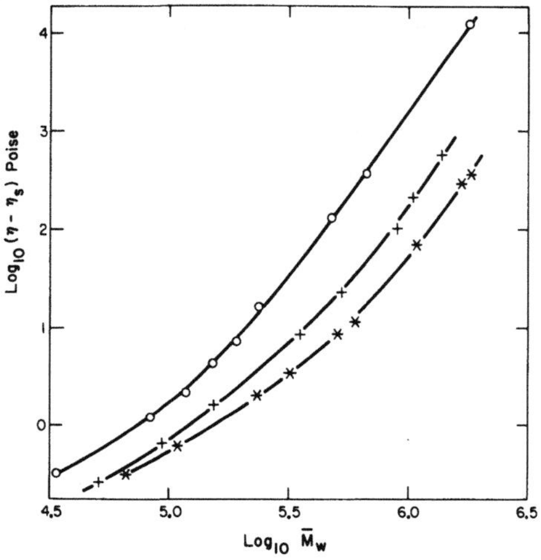

Fig. 11. Viscosities of linear and branched polystyrenes in diethylbenzene (0.255 g/ml) at 30°. Symbols as in Fig. 8, Ref. [74]

ment (viscosities much higher than expected from simple consideration of coil size) at a concentration of 0.33 g/ml[75] (Fig. 12).

No entirely satisfactory description of the visco-elastic properties of concentrated solutions of even linear polymers is available at this time[78] so only simplified mechanisms of viscosity enhancement caused by branching can be expected. The diffusion of a linear flexible macromolecule in a medium containing loop like obstacles caused by entanglements from other polymer chains has been represented as a snaking motion in a "tunnel" created by the loops[79]. The tunnel constantly reorganizes to fit the changing chain contour and may well also diffuse since it is composed of transient entanglement points. A branched polymer would create a branched tunnel, if there are enough entanglement points per branch, which could render this type of motion (the circulatory type described by Bueche) much more difficult if the branches are long. A high resistance to motion occurs at the branch point[75]. A simplified geometrical model of diffusion in a long tube has been used[75] to calculate the resistance to flow (V), the size of a side tunnel being regarded as critical. This leads to the expression

$$V = Z\,\phi^{5/6}\left[\frac{\phi M_{br}}{M_c}\right] \tag{17}$$

where Z is a parameter calculated from physical data on the polymer, which varies from 0.08 for polystyrene to 0.10 for polyisoprene and to 0.13 for polybutadiene i.e. not greatly dependent on polymer structure. ϕ is the volume fraction of polymer, M_{br} is the branch molecular weight and M_c the characteristic molecular weight for the onset of entangled flow in bulk polymer. M_c is known to vary more with polymer constitution, being ~32,000 for polystyrene ~10,000 for polyisoprene and ~6000 for polybutadiene. The calculated effect is supposed to be superposed on the normal factors influencing flow in linear systems, so that it should correlate with viscosity enhancement. Indeed the enhancement in bulk polybutadiene and diluted polyisoprene were found to be a single function of the factor $\phi^{5/6}\left[\frac{\phi M_{br}}{M_c}\right]$[75)]

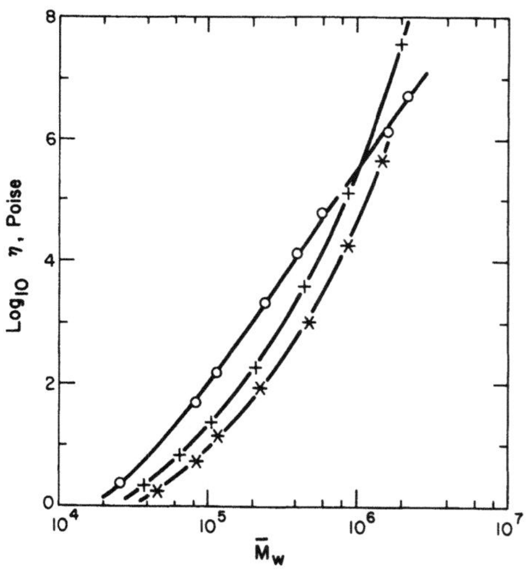

Fig. 12. Viscosities of linear and branched polyisoprenes in tetradecane (0.33 g/ml) at 25°. Symbols as in Fig. 8, Ref. [75]

for four-branch star polymers. Inclusion of the Z factor worsened the agreement unless the ϕ term was adjusted. Nevertheless the model appears to rationalize the results on branched polymers satisfactorily.

Viscosity enhancement in branched polystyrenes would be expected to be considerably smaller than in branched diene polymers. A smaller Z factor and a higher M_c for this polymer combine to reduce the V value. The star-branched polymers shown in Fig. 11 would not therefore be expected to show viscosity enhancement at a concentration of 0.25 g/ml. Experiments reported[80] on melt viscosities of star-branched polystyrenes having $f \approx 3$ on the other hand would have been expected to have shown some enhancement at least at the highest molecular weight. None was in fact found, the viscosities being approximately predictable from the Bueche theory i. e. were all lower than their linear counterparts. Viscosity enhancements have been reported for multibranch star polystyrenes (f = 7 to 13)[81] in melt flow experiments. These are of comparable magnitude to the values found for four-branch star polybutadienes at the same value of M_{br}/M_c. Specific effects of multiple branching were not considered in the model described above but there is evidence that the major enhancement effect is produced at f = 3. Increase of the branch number to four at constant M_{br} increases the enhancement by about a factor of two but subsequent increases in f have only small effect[81].

Spring-bead models of viscoelasticity of flexible polymers predict decreases in the steady state compliance J_e^0 on branching either in free draining or non-free draining versions. A decrease in J_e^0 at infinite dilution is in fact found as noted earlier. In concentrated solutions the Rouse theory still successfully predicts the viscosity behaviour ($\eta \sim M$) at low molecular weights despite overlapping of polymer domains. It fails for higher molecular weights where entanglements are prevalent, both for linear and branched polymers. Parallel changes occur in other viscoelastic parameters. J_e^0, predicted to be $0.4\ g_2 M/cRT$ on the Rouse theory, tends to become independent of molecular weight and increasingly long plateaus occur in the shear modulus versus frequency plots as the molecular weight is increased[82]. The transient entanglement network must profoundly influence the spectrum of relaxation times. The longer relaxation time distribution, sensitive to large scale chain motions, sharpens and

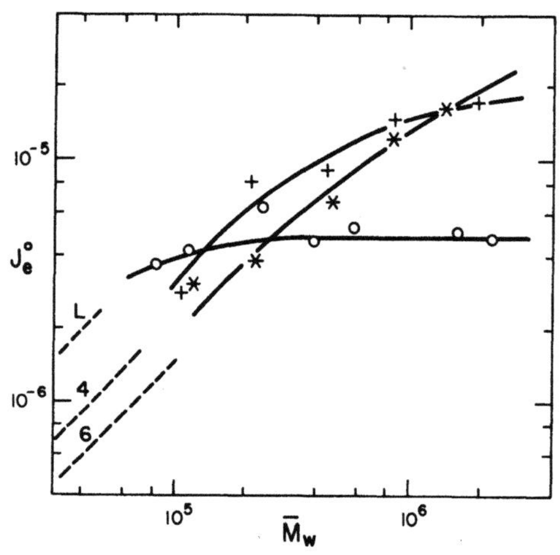

Fig. 13. Steady state compliance of linear and branched polyisoprenes in tetradecane (0.33 g/ml) at 25°. Symbols as in Fig. 8, Ref. [75]. Dashed lines are values calculated from Rouse-Ham Theory

moves to longer times, separating from the faster relaxations which may retain Rouse-like characteristics.

Various models have been suggested to predict the effects of entanglements on visco-elastic properties of linear polymers including some which simply modify the Rouse theory. For a review of this topic the reader is referred to a full account in an earlier volume of this series[78]. None is completely satisfactory in explaining the whole range of viscoelastic effects so that a meaningful treatment of the effects of branching is not possible at this time. Figure 13 shows the steady state compliance of concentrated solutions of polyisoprene. The near independence of J_e^0 of the linear polymers is evident. J_e^0 for the branched polymers attains values higher than those of linear polymers and appears to approach higher plateau values as branching increases. All tend towards the normal values predicted by the Rouse-Ham theory at low molecular weights i.e. in the range where Rouse-like viscosity behaviour is observed. Indeed the compliance of the branched polymers appears to be dependent on molecular weight over a much wider range before approaching a limiting value.

These branched polyisoprenes exhibited viscosity enhancement (Fig. 12) but this does not seem to be a prerequisite for high values of J_e^0. Measurements on star-branched polystyrenes[80] in the melt also indicated high J_e^0 values whereas the viscosities were lower than those of linear polymers. The same trend in J_e^0 is evident[83] for the star-branched polystyrenes illustrated in Fig. 11. The plateau modulus of the star-branched polystyrenes on the other hand was virtually unchanged between linear and branched samples[80]. As this can be related to the molecular weight between entanglement points, this factor seems to be unchanged on branching and hence unlikely to be a source of changes in J_e^0. It appears that compliance enhancement in branched polymers is connected with a broadening of the distribution of longer relaxation times[75].

The viscosities referred to so far have been those measured at zero shear. Experiments have been reported[75, 76, 83] on the dependence of viscosity on shear rate. The form of the viscosity/shear rate function appears to be the same for linear and star-branched polymers if both are reasonably monodisperse. At a fixed molecular weight, higher shear rates are required to observe similar deviations in η (e.g. a drop of η to 80% of its initial value) with branched polymers under conditions where no viscosity enhancement occurs. Taken at fixed radius of gyration deviations in η occur at about the same shear rate. With higher molecular weight star-branched polymers the onset of shear rate dependence can occur at lower shear rates than for linear polymers i.e. both viscosity and its shear dependence can be enhanced. Under these conditions a star-branched polymer can show a higher viscosity at low shear rates than its linear counterpart of the same radius of gyration but a lower viscosity at high shear rates.

5 The Glass Transition

In general the properties of the glassy state would be expected to be unchanged on branching, at least until very low molecular weights, where the differences in molec-

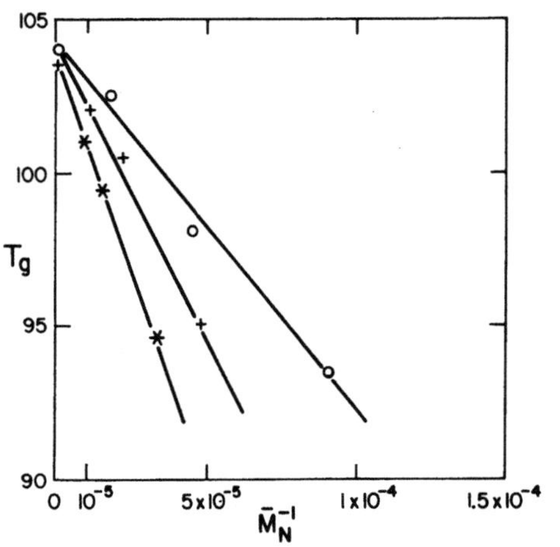

Fig. 14. Effect of molecular weight on T_g of linear and branched polystyrenes. Symbols as in Fig. 8, Ref.[84] (penetration method)

ular volume of the branch points and the chain ends from that of the segments could become important. In a study of four and six-branched polystyrenes with molecular weights as low as 20,000[84] the specific volume and expansion coefficient below T_g were essentially unchanged from the values appropriate to high molecular weight linear polymer. Above T_g, expansion coefficients were similarly identical. Depression of T_g below the value appropriate to linear high molecular weight polymers was observed for the lower molecular weight star-branched polymers. Figure 14 shows that this effect is larger at a given molecular weight the larger is the number of branches. The phenomenon of T_g depression for linear polymers of low molecular weight is a well-known phenomenon and generally attributed to the disruptive effects of chain ends on packing of the polymer chains above T_g, leading to enhanced values of specific volume and hence free volume. Since branched polymers have for a given molecular weight a larger number of chain ends, a larger effect in this case is not unexpected. Meares[85] has suggested that the conventional relationship between T_g and $\bar{M}_n$ should be modified to

$$T_g = T_g^{\infty} - \frac{K\omega}{2\,\bar{M}_n} \tag{18}$$

where ω is the number of chain ends per molecule and T_g^{∞} is the value appropriate to high molecular weight polymer. This equation reduces to the form for linear polymers described by Fox and Flory [86] when $\omega = 2$. The results for linear, four and six-branched star polystyrenes conform to this relationship[84]. Under appropriate conditions it is conceivable that the central branch point could also have an effect analogous to that observed in polymer networks, where T_g is increased by the presence of the more compact branch points. No evidence for this effect could be obtained for polymers with branch lengths down to M = 5000 but it is possible that at very low chain lengths the chain end and branch point effects could partially compensate.

Acknowledgement. The author wishes to thank Dr. J. E. L. Roovers for providing much unpublished data.

6 References

1. Schaefgren, J. R., Flory, P. J.: J. Am. Chem. Soc. *70*, 2709 (1948)
2. Pfannemüller, B., Burchard, W.: Makromol. Chem. *121*, 1 (1969)
3. Burchard, W., Eschwey, A., Franken, I., Pfannemüller, B.: Structure of fibrous biopolymers. Colston Papers *26*, 365 (1974)
4. Pfannenmüller, B. Burchard, W., Franken, I.: Stärke *28*, 1 (1976)
5. Swarc, M.: Carbanions, living polymers and electron transfer processes. New York: Interscience 1968
6. Bywater, S.: Progr. Polym. Sci. *4*, 27 (1974)
7. Bywater, S.: Comprehensive chemical kinetics. Amsterdam: Elsevier *15*, 1 (1976)
8. Morton, M., Helminiak, T. E., Gadkary, S. D., Bueche, F.: J. Polym. Sci. *57*, 471 (1962)
9. Wenger, F., Yen, S.-P. S.: Polymer Reprints A.C.S. *3* (1) 162 (1962)
10. Altares, T., Wyman, D. P., Allen, V. R., Meyerson, K.: J. Polym. Sci. (Part A) *3*, 4131 (1965)
11. Meunier, J.-C., van Leemput, R.: Makromol. Chem. *142*, 1 (1971)
12. Gervasi, J. A., Gosnell, A. B.: J. Polym. Sci. (Part A1) *4*, 1391 (1966)
13. Zelinsky, R. P., Wofford, C. F.: J. Polym. Sci. (Part A) *3*, 93 (1965)
14. Hsieh, H. L.: Rubber Chem. Tech. *49*, 1307 (1976)
15. Roovers, J. E. L., Bywater, S.: Macromolecules *5*, 384 (1972)
16. Herz, J., Strazielle, C.: C. R. Acad. Sci. *272*, 747 (1971)
17. Herz, J., Hert, M., Strazielle, C.: Makromol. Chem. *160*, 213 (1972)
18. Hert, M., Strazielle, C., Herz, J.: C. R. Acad. Sci. *C276*, 395 (1973)
19. Bryce, W. A. J., McGibbon, G., Meldrum, J. G.: Polymer *11*, 394 (1970)
20. Decker, D., Rempp, P.: C. R. Acad. Sci. *261*, 1977 (1965)
21. Zilliox, J.-G., Decker, D., Rempp, P.: C. R. Acad. Sci. *262*, 726 (1966)
22. Zilliox, J.-G., Rempp, P., Parrod, J.: J. Polym. Sci. (Part C) *22*, 145 (1968)
23. Worsfold, D. J., Zilliox, J.-G., Rempp, P.: Can. J. Chem. *47*, 3379 (1969)
24. Worsfold, D. J.: Macromolecules *3*, 514 (1970)
25. Eschwey, H., Burchard, W.: Polymer *16*, 180 (1975)
26. Kohler, A., Zilliox, J.-G., Rempp, P.: Eur. Polym. J. *8*, 627 (1972)
27. Eschwey, H., Hallensleben, M. L., Burchard, W.: Makromol. Chem. *173*, 235 (1973)
28. Bi, L.-K., Fetters, L. J.: Macromolecules *9*, 732 (1976)
29. Flory, P. J.: Principles of polymer chemistry. Ithaca: Cornell U. Press, 1953.
30. Zimm, B. H., Stockmayer, W. H.: J. Chem. Phys. *17*, 1301 (1949)
31. Burchard, W.: Macromolecules *7*, 835, 840 (1974)
32. Burchard, W.: Macromolecules *10*, 919 (1977)
33. Burchard, W.: British Polym. J. *9*, 103 (1977)
34. Benoit, H.: J. Polym. Sci. *11*, 507 (1953)
35. Stockmayer, W. H., Fixman, M.: Ann. N.Y. Acad. Sci. *57*, 334 (1953)
36. Šolc, K.: Macromolecules *6*, 378 (1973)
37. Thurmond, C. D., Zimm, B. H.: J. Polym. Sci. *8*, 477 (1952)
38. Rouse, P. E.: J. Chem. Phys. *21*, 1272 (1953)
39. Zimm, B. H.: J. Chem. Phys. *24*, 269 (1956)
40. Zimm, B. H., Kilb, R. W.: J. Polym. Sci. *37*, 19 (1959)
41. Ham, J. S.: J. Chem. Phys. *26*, 625 (1957)
42. Osaki, K., Schrag, J. L.: J. Polym. Sci. (Phys. Ed.) *11*, 549 (1973)
43. Osaki, K.: Adv. Polymer Sci. *12*, 1 (1973)
44. Bueche, F.: J. Chem. Phys. *20*, 1959 (1952)
45. Bueche, F.: J. Chem. Phys. *20*, 599 (1952)
46. Berry, G. C., Fox, T. G.: Adv. Polymer Sci. *5*, 261 (1968)
47. Bueche, F.: J. Chem. Phys. *40*, 484 (1964)
48. Orofino, T. A., Wenger, F.: J. Phys. Chem. *67*, 566 (1963)
49. Roovers, J. E. L., Bywater, S.: Macromolecules *7*, 443 (1974)
50. Berry, G. C.: J. Polym. Sci. (Part A2) *9*, 687 (1971)

51. Strazielle, C., Herz, J.: Eur. Polym. J. *13*, 223 (1977)
52. Meunier, J.-C., van Leemput, R.: Makromol. Chem. *147*, 191 (1971)
53. Zilliox, J.-G.: Makromol. Chem. *156*, 121 (1972)
54. Hadjichristidis, N., Roovers, J. E. L.: J. Polym. Sci. (Phys. Ed.) *12*, 2521 (1974)
55. Candau, F., Rempp, P., Benoit, H.: Macromolecules *5*, 627 (1972)
56. Mazur, J., McCrackin, F.: Macromolecules *10*, 326 (1977)
57. Toporowski, P. M., Roovers, J. E. L.: Macromolecules *11*, 365 (1978)
58. Orofino, T. A., Flory, P. J.: J. Chem. Phys. *26*, 1067 (1957)
59. Candau, F., Strazielle, C., Benoit, H.: Makromol. Chem. *170*, 165 (1973)
60. Stockmayer, W. H.: J. Polym. Sci. *15*, 595 (1955)
61. Berry, C. G., Orofino, T. A.: J. Chem. Phys. *40*, 1614 (1964)
62. Kurata, M., Fukatsu, M.: J. Chem. Phys. *41*, 2934 (1964)
63. Osaki, K., Mitsuda, Y., Johnson, R. M., Schrag, J. L., Ferry, J. D.: Macromolecules *5*, 17 (1972)
64. Mitsuda, Y., Osaki, K., Schrag, J. L., Ferry, J. D.: Polym. J. *4*, 24 (1973)
65. Roovers, J. E. L.: unpublished
66. Berry, G. C.: J. Chem. Phys. *44*, 4550 (1966)
67. Yamakawa, H.: Modern theory of polymer solutions. New York: Harper and Row 1971
68. Grubisic, Z., Rempp, P., Benoit, H.: J. Polym. Sci. (Part B) *5*, 753 (1967)
69. Benoit, H., Grubisic, Z., Rempp, P., Decker, D., Zilliox, J.-G.: J. Chim. Phys. *63*, 1507 (1966)
70. Casassa, E. F., Tagami, Y.: Macromolecules *2*, 14 (1969)
71. Roovers, J. E. L., Hadjichristidis, N.: unpublished
72. Pannell, J.: Polymer *13*, 277 (1972)
73. Kato, T., Itsubo, A., Yamamoto, Y., Fujimoto, T., Nagasawa, M.: Polym. J. *7*, 123 (1975)
74. Utracki, L. A., Roovers, J. E. L.: Macromolecules *6*, 366 (1973)
75. Graessley, W. W., Masuda, T., Roovers, J. E. L., Hadjichristidis, N.: Macromolecules *9*, 127 (1976)
76. Kraus, G., Gruver, J. T.: J. Polym. Sci. (Part A) *3*, 105 (1965)
77. Kraus, G., Gruver, J. T.: J. Polym. Sci. (Part A 2) *8*, 305 (1970)
78. Graessley, W. W.: Adv. Polymer Sci. *16*, 1 (1974)
79. de Gennes, P.: J. Chem. Phys. *55*, 572 (1971)
80. Masuda, T., Ohta, Y., Onogi, S.: Macromolecules *4*, 763 (1971)
81. Graessley, W. W.: Accounts Chem. Res. *10*, 332 (1977)
82. Ferry, J. D.: Viscoelastic properties of polymers. (2nd edit.). New York: Wiley 1970
83. Utracki, L. A., Roovers, J. E. L.: Macromolecules *6*, 373 (1973)
84. Roovers, J. E. L., Toporowski, P. M.: J. Appl. Polym. Sci. *18*, 1685 (1974)
85. Meares, P.: Trans. Faraday Soc. *53*, 31 (1957)
86. Fox, T. G., Flory, P. J.: J. Appl. Phys. *21*, 581 (1950)

Received July 27, 1978
H.-J. Cantow (editor)

Dilute Solution Properties of Aliphatic Polyamides

Z. Tuzar, P. Kratochvíl, and M. Bohdanecký

Institute of Macromolecular Chemistry,
Czechoslovak Academy of Sciences, 162 06 Prague 6, CSSR

The development of and new trends in the determination of molecular weights and molecular weight distribution of aliphatic unsubstituted linear polyamides (nylons) are reviewed here. Critically discussed are intrinsic viscosity – molecular weight relationships and the conformational regidity of nylon chains. Attention is also given to some peculiarities in the behavior of nylons in multicomponent solvents.

Table of Contents

1 Introduction

In the rich assortment of polyamides an important position is held by aliphatic unsubstituted linear polyamides, known under the current name of nylons. They are prepared either by polycondensation of aliphatic diacids and diamines or of aliphatic ω-amino acids, or by polyaddition of their lactams. The numeral(s) following the term nylon indicate(s) the number of carbon atoms in the monomeric unit: thus, *e. g.* the polymer of pyrrolidone is called nylon 4, that of caprolactam is nylon 6, the copolycondensate of hexamethylenediamine and adipic acid is nylon 6,6 and the copolycondensate of hexamethylenediamine and sebacic acid is nylon 6,10.

For several decades, nylons 6 and 6,6 have been virtually the only nylons widely used on an industrial scale. It was not until recently that some other members of the nylon series – from nylon 4 to nylon 12 – became an object of increased interest.

There are a considerable number of papers dealing with dilute nylon solutions, especially those of nylon 6 and nylon 6,6 and the disagreement between results of different authors is conspicuous. The main reasons are, on the one hand, the obsoleteness of some papers, the disputable findings of which have been uncritically accepted up to the present day, and on the other hand, the peculiarities of the solution behavior of nylons. The molecular characterization of nylons, *e. g.* determination of molecular weights and dimensions, may in many solvents be complicated by the polyelectrolyte effect, the presence of aggregates, hydrolysis, or by phenomena ensuing from the complicated behavior of polymer solutions in multicomponent solvents.

This article is an attempt at a comprehensive survey of papers dealing with dilute solutions of nylons, especially of nylons 6, 6,6, 4 and 12. The individual chapters will be devoted to the determinations of molecular weights and molecular weight distribution, to the hydrodynamic properties and conformation of nylon molecules, and finally to some specific features of the behavior of nylons in multicomponent solvents. Since the solution properties of these nylons are alike, the individual nylons are not dealt with separately; differences in the behavior of various nylons are pointed out in the respective chapters.

Since nylons prepared by various procedures, *e. g.* by the hydrolytic or alkaline polymerization, do not generally differ much either in molecular structure or in dilute solution properties, the method of preparation is not mentioned in the text. The polymerization method is pointed out only in those cases where side reactions affect the structure of the product (*e. g.*, branching), or effects due to the nature of various end groups are operative.

2 Solvents

The behavior of nylons in solution is strongly affected by the presence of polar amide groups –CONH–. Perfect dissolution of nylon requires such a strong interaction between the solvent and the amide group that interamidic hydrogen bonds in the polymer would be destroyed. An attempt to classify solvents of nylon 6,6, involving also compounds that are solvents at elevated temperature, has been described in Ref.[1]. The usual solvents of nylon 6 and 6,6 are traditionally divided into four groups[2]: (a) strong acids (*e. g.* formic acid, sulphuric acid), (b) phenols (*e. g.* phenol, cresol), (c) aliphatic fluorinated alcohols (*e. g.* 2,2,3,3-tetrafluoropropanol, 2,2,2-trifluoroethanol), (d) concentrated solutions of some salts in alcohols.

The solubility of the other nylons predominantly depends on the frequency of amide groups in the chain. While nylon 4 with a higher frequency of amide groups can be dissolved in all virtually solvents of nylon 6 and 6,6, the assortment of solvents of, *e. g.* nylon 12, is poorer owing to its longer methylene sequences. The effect of frequency of the amide groups on solubility can be adequately documented by the solubility of various nylons in formic acid[3] (Fig. 1). In the series from nylon 4 to nylon 12 dissolution requires increasingly more and more concentrated acid. The concentration range of formic acid between curves a) and b) in Fig. 1, which decreases with decreasing frequency of the amide groups, may be explained as follows: At a given temperature (25 °C in our case), the hydrogen bonds between the amide groups of nylon may be destroyed only in an acid up to a certain thermodynamic quality (given by the water content – curve a)). Within a very restricted range below this limit, the nylon may be dissolved (*i. e.* interamidic hydrogen bonds may be destroyed) either by heating, or within a broader concentration range of the acid by adding water to a solution in a more concentrated acid (curve b) – Fig. 1). In solutions in this range aggregation or even precipitation of nylon takes place, depending on the molecular weight and concentration of

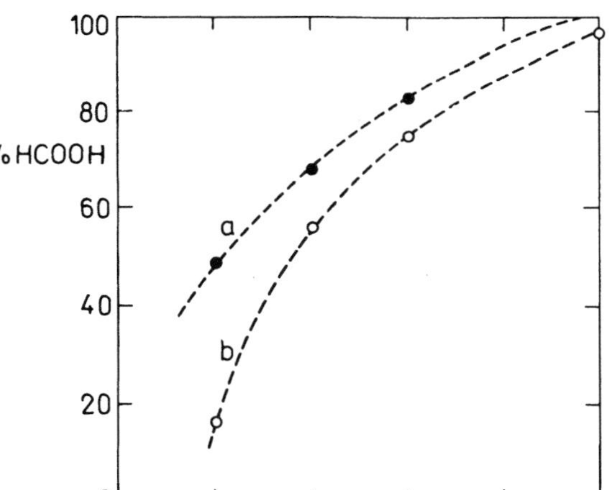

Fig. 1. Solubility of nylons 4, 6, 8 and 12 in HCOOH at 25 °C; a) concentrations of HCOOH (wt.-%) at which nylons are directly soluble, b) concentrations of HCOOH (wt.-%) at which nylons directly precipitate if water is added[3]

nylon and acid. Nylon 12 may be dissolved only in a 98–100 % formic acid and only at elevated temperatures. At 25 °C it precipitates from solution.

The effect of frequency of amide groups may also be observed at dissolution of nylons in the mixture 2,2,3,3-tetrafluoropropanol/H_2O: Nylon 4 is dissolved in a mixture with 54 vol.-% H_2O[4], nylon 6 is dissolved in a mixture with 32 vol.-% H_2O[5], nylon 8 is dissolved in a mixture with 10 vol.-% H_2O, while nylon 12 does not dissolve at room temperature even in anhydrous 2,2,3,3-tretrafluoropropanol[6].

Difficulties arise due to the hydrolytic cleavage of amide bonds in strong acids. Statistical hydrolytic cleavage of nylon 6 and 6,6 has been described in hydrochloric acid[7], sulphuric acid[8–13] and formic acid[10, 13, 14]. The relative decrease in molecular weight in a given acid is proportional to the number of amide groups in the chain; for one kind of nylon it increases with molecular weight, while for different nylons it increases with the frequency of amide groups in the chain[6]. For this reason, the use of strong acids as solvents of nylon 4, and sometimes of nylon 6, may be risky.

While according to Valentine's classification[2] the first three groups of solvents mentioned above include compounds of acid character, *i. e.* they contain either a carboxyl or a hydroxyl group with an acid hydrogen atom, the fourth group involves solutions of salts, whose cations are able to form relatively strong complexes with the nylon amide group (See Chapter 6). Owing to such specific solvation mechanisms, these solvents more effectively dissolve nylons with a higher frequency of amide bonds, while nylon 12 for example, remains undissolved.

There are only a few one-component solvents of nylons, *e. g.* the universal solvent of all nylons – cresol, or some fluorinated alcohols. Multicomponent solvents are the ones mostly used in the study of solution properties, mainly in the determination of the molecular parameters of nylons. The main reason for this is the specific demands of various methods on the refractive index, viscosity, density, boiling point and other properties, which are difficult to meet by means of one-component solvents because of their restricted number. Other reasons leading to the use of multicomponent solvents are, *e. g.* an effort to suppress the polyelectrolyte effect (See 5.1), to avoid aggregation or to modify the thermodynamic quality of the solvent (See 3.2).

Particular solvent systems for various nylons are dealt with in the individual chapters; some specific features of multicomponent solvents are discussed in Chapter 6.

3 Methods of Molecular Weight Determination

The number-average molecular weight, M_n, of nylons is most frequently determined by end group analysis; less frequently, colligative methods, such as cryoscopy, vapour pressure osmometry (VPO), or membrane osmometry are used. With minor exceptions, the weight-average molecular weight, M_w, of nylons is determined by light scattering. The methods themselves are not described in greater detail, since competent reviews[15–17] are available.

3.1 End Group Analysis

The molecules of nylons are usually terminated with the carboxyl and amine groups. The analytical determination of these groups ranks among the oldest methods of molecular weight measurements of nylons. From the first attempts by Carothers[18] up to very sensitive methods suggested by Garmon and Gibson[19], end group analysis has been increasingly improved with the development of analytical methods as regards its reliability, accuracy and requirements on the amount of the sample. Of course, the accuracy of determination increases with decreasing molecular weight. While the upper limit for a reliable determination of M_n was earlier given as 25×10^3, it is now quite common to determine molecular weights of commercial nylons up to about 50×10^3 with sufficient accuracy. The determination of M_n of nylons by end group analysis has been reviewed several times[16, 20].

Reagents most frequently used in titration of the amine and carboxyl groups are hydrochloric acid and potassium hydroxide; other reagents, such as *e. g.* perchloric acid[21,22] or *p*-toluenesulphonic acid[23], are seldom employed. Indication is visual[24–27], potentiometric[21, 22, 28], or conductometric[24, 28, 29]. The tritration medium is selected according to the character of the sample and the type of indication. A number of solvent systems have been described in the literature. In titration of the amine groups, the solvent systems are *e. g.* phenol with aqueous ethanol[24], with propanol[30] or with glycol[22], cresol with methanol[27], with isopropanol[28] or with acetic acid [21], benzyl alcohol[25], sometimes with aqueous methanol[26], and 2,2,2-trifluoroethanol[23]. Carboxyl groups are determined in *e. g.* benzyl alcohol[24], sometimes with aqueous ethanol[26] or with propanol[30], in propargyl alcohol[28] or in a mixture of 2,6-xylenol with chloroform and acetone[31].

The titrimetric determination of amine and carboxyl groups is very sensitive to the presence of extraneous bases and acids. Since samples are usually fractionated or at least reprecipitated before measurement, often in an acid medium, the traces of acid not removed by washing are the most frequent source of errors. Garmon and Gibson have developed a method free of this kind of error[19]. They titrated the amine groups with 2,4-dinitrofluorobenzene, which is insensitive to strong bases; carboxyl groups were esterified with labelled methanol (^{14}C) and determined by the scintillation method. The results were practically identical with those obtained by the most frequently used classical titrimetric method of Waltz and Taylor[24].

In order to regulate the molecular weight, acetic acid, which forms practically quantitatively nonreactive N-acetyl end groups, is frequently added to the polymerizing lactam. After hydrolysis these groups can be determined titrimetrically[32] or colorimetrically[33] as acetic acid.

The carboxyl groups may also be esterified with diazomethane and the methyl ester groups may then be determined[34, 35].

Errors involved in the determination of M_n by end group analysis, which may sometimes be large, are not of methodical character in most cases. Usually, they are due to the lower or higher concentration of the amine and carboxyl end groups compared with the assumed value corresponding to the consumption of the initiator. In the case of decay of the titratable groups owing to side reactions[36] or aging[37], the method yields higher M_n, while branching decreases M_n. Distorted results may be obtained in the determination of end groups of relatively low-molecular weight nylons (*e. g.*, low-molecular weight extracts[1]), because the method cannot be adequately applied to cyclic oligomers.

3.2 Colligative Methods

In the determination of M_n lower than about 10^4, only cryoscopy, ebulliometry and vapor pressure osmometry (VPO) can compete with the end group method. In practice, nylons with a higher M_n are produced, while samples with $M_n < 10^4$ are met only rarely, *e. g.* as oligomeric extracts[1]. This is why these methods are seldom applied to nylons.

Cryoscopy was used only in preliminary measurements of nylon 6,6 ($M_n \approx 3 \times 10^4$)[38] and in the determination of a number of fractions of mixed nylon 6 and 6,6 ($M_n \approx (1-5) \times 10^3$) in phenol[39].

The VPO method consists in measurement of the temperature difference between a drop of solution and solvent in the stationary state in the saturated solvent vapor. Solvents should possess a high heat of evaporation, moderate vapor pressure and low viscosity. Plausible results were obtained with nylon 6 in 90 wt.-% formic acid at 37 °C[40]. Owing to the strong corrosiveness of the solvent, the apparatus had to be adapted by gilding the vapor cell. In further papers various authors have attempted to circumvent this disadvantage by using less corrosive solvents, such as 2,2,3,3-tetrafluoropropanol for nylon 6 at 20 °C[41], the same solvent at 50 °C for nylons 6, 7 and 9[42], or 2,2,2-trifluoroethanol and *o*-chlorophenol at 37 °C for nylon 6,6[1].

A colligative method of the determination of M_n most frequently used with M_n ranging between 1×10^4 and 1×10^6 is membrane osmometry. In the osmometry of nylons an important role is played by the choice of solvent and material of the membrane. Although difficulties involved in the application of cellulose membranes in formic acid were not mentioned[43], later on glass membranes were given preference in this solvent[44]. In one case, *i. e.* in a cresol/toluene mixture, collodium membranes were employed[45]. For a number of solvents (phenol/methanol[39], o-chlorophenol[1, 46], 2,2,3,3-tetrafluoropropanol with sodium trifluoroacetate[1]), common cellulose membranes seem to be adequate. The establishment of osmotic equilibrium with more viscous solvents may be accelerated by raising the temperature; 60 °C had to be used with cresol solutions[44, 47], while 40 °C and 60 °C were the respective values for 2,2,3,3-tetrafluoropropanol with trifluoroacetate[1] and for o-chlorophenol[46]. In an extensive review[1], o-chlorophenol was recommended out of a large number of potential solvents for the osmometry of nylons.

Table 1. Refractive index increments of nylons

Solvent	dn/dc (ml/g)		°C	Ref.
	436 nm	546 nm		
Nylon 4				
3 g $CaCl_2$/10 ml (H_2O C_2H_5OH)/35 vol.-%		0.177[a]	25	122)
m-Cresol		0.015	25	62)
m-Cresol/10 vol.-% n-heptane		0.042[a]	25	62)
m-Cresol/20 vol.-% n-heptane		0.121[a]	25	62)
m-Cresol/30 vol.-% n-heptane		0.234[a]	25	62)
m-Cresol/40 vol.-% n-heptane		0.415[a]	25	62)
m-Cresol/50 vol.-% n-heptane		0.710[a]	25	62)
TFP		0.202	25	4)
TFP/15 vol.-% H_2O		0.192[a]	25	4)
TFP/35 vol.-% H_2O		0.181[a]	25	4)
TFP/50 vol.-% H_2O		0.155[a]	25	4)
Nylon 6				
Chloralhydrate	0.075	0.072	60	48)
m-Cresol		0.005	25	62)
m-Cresol/10 vol.-% n-heptane		0.047[a]	25	62)
m-Cresol/20 vol.-% n-heptane		0.132[a]	25	62)
m-Cresol/30 vol.-% n-heptane		0.243[a]	25	62)
m-Cresol/40 vol.-% n-heptane		0.471[a]	25	62)
m-Cresol/50 vol.-% n-heptane		0.816[a]	25	62)
51 wt.-% H_2SO_4		0.106	25	13)
46 wt.-% H_2SO_4		0.116	25	13)
44 wt.-% H_2SO_4		0.119	25	13)
42 wt.-% H_2SO_4		0.122	25	13)
35.5 wt.-% H_2SO_4		0.129	25	13)
33 wt.-% H_2SO_4		0.131	25	13)
TFP/0.1M LiCl/10 vol.-% H_2O		0.182[a]	25	5)
TFE		0.220	25	46)
TFE	0.263		25	58)
TFE		0.236	25	138)
Nylon 6,6				
m-Cresol		−0.016	25	27)
Dichloroacetic acid		0.098	25	27)

Table 1 (continued)

Solvent	dn/dc (ml/g)		°C	Ref.
	436 nm	546 nm		
Dichloroacetic acid		0.099	50	27)
Dichloroacetic acid		0.104	80	27)
100 wt.-% HCOOH	0.157		25	54)
95 wt.-% HCOOH	0.150		25	54)
90 wt.-% HCOOH	0.145		25	54)
90 wt.-% HCOOH		0.145	25	27)
90 wt.-% HCOOH		0.142	25	146)
85 wt.-% HCOOH	0.141		25	51)
80 wt.-% HCOOH	0.145		25	54)
75 wt.-% HCOOH	0.144		25	54)
100 wt.-% HCOOH/ 0.5M HCOONa	0.136		25	54)
95 wt.-% HCOOH/ 0.5M HCOONa	0.136		25	54)
90 wt.-% HCOOH/ 0.5M HCOONa	0.136		25	54)
90 wt.-% HCOOH/ 0.5M HCOONa	0.137		25	54)
80 wt.-% HCOOH/ 0.5M HCOONa	0.136		25	54)
75 wt.-% HCOOH/ 0.5M HCOONa	0.138		25	54)
90 wt.-% HCOOH/ 0.025M HCOONa	0.147		25	54)
90 wt.-% HCOOH/ 0.05M HCOONa	0.146		25	54)
90 wt.-% HCOOH/ 0.1M HCOONa	0.142		25	54)
90 wt.-% HCOOH/ 0.2M HCOONa	0.142		25	54)
90 wt.-% HCOOH/ 0.75M HCOONa	0.130		25	54)
90 wt.-% HCOOH/ 1.0M HCOONa	0.124		25	54)
90 wt.-% HCOOH/ 0.2M KCl	0.143		25	53)
90 wt.-% HCOOH/ 0.5M KCl	0.140		25	53)
90 wt.-% HCOOH/ 1.0M KCl	0.136		25	53)
90 wt.-% HCOOH/ 1.5M KCl	0.131		25	53)
90 wt.-% HCOOH/ 2.5M KCl	0.122		25	53)
98 wt.-% HCOOH/ 2M KCl	0.137		25	53)
95 wt.-% HCOOH/ 2M KCl	0.129		25	53)

Table 1 (continued)

Solvent	dn/dc (ml/g) 436 nm	dn/dc (ml/g) 546 nm	°C	Ref.
90 wt.-% HCOOH/ 2M KCl	0.126		25	53)
85 wt.-% HCOOH/ 2M KCl	0.124		25	53)
82.5 wt.-% HCOOH/ 2M KCl	0.123		25	53)
90 wt.-% HCOOH/ 2M KCl		0.126	25	27)
Octafluoropentanol		0.192	25	27)
95 wt.-% H_2SO_4		0.081	25	27)
TFP/0.1M NaTFAc	0.190		25	49)
TFE		0.228	25	138)
Nylon 6,10				
90 wt.-% HCOOH		0.156	25	144)
Nylon 12				
m-Cresol		−0.020	25	62)
m-Cresol/10 vol.-% n-heptane		0.036[a]	25	62)
m-Cresol/20 vol.-% n-heptane		0.075[a]	25	62)
m-Cresol/30 vol.-% n-heptane		0.182[a]	25	62)
m-Cresol/40 vol.-% n-heptane		0.340[a]	25	62)
m-Cresol/20 vol.-% TFP		0.060[a]	25	60)
m-Cresol/30 vol.-% TFP		0.105[a]	25	60)
m-Cresol/40 vol.-% TFP		0.165[a]	25	60)
m-Cresol/50 vol.-% TFP		0.205[a]	25	60)
m-Cresol/60 vol.-% TFP		0.254[a]	25	60)
m-Cresol/65 vol.-% TFP		0.285[a]	25	60)
HFP	0.266	0.255	25	59)

TFP 2,2,3,3−tetrafluoropropanol, TFE 2,2,2-trifluoroethanol, NaTFAc sodium trifluoroacetate, HFP 1,1,1,3,3,3-hexafluoro-2-propanol.

[a] values obtained after dialysis of nylon solution against the given solvent.

3.3 Light Scattering

With small exceptions, light scattering is virtually the only method used in the determination of the weight-average molecular weight, M_w, of nylons. In addition to molecular weight, the method supplies data on the thermodynamic quality of the solvent (second virial coefficient A_2); for higher M_w ($>10^5$), it also provides information on the dimensions of polymer molecules (radius of gyration). The shape of the angular dependence of scattered light may indicate the presence of even a tiny quantity of aggregates or microgels.

The choice of solvents for measurements of M_w by light scattering is restricted by the following requirements: (a) in the given solvent nylon must possess a sufficiently high refractive index increment dn/dc (the published dn/dc values are summarized in Table 1); (b) solutions must be molecular (*i. e.* free from aggregates) and stable in time; (c) solutions must not exhibit the polyelectrolyte effect (See 5.1).

If the M_w of nylons is to be determined within the usual range (the lower limit of M_w being about 1×10^4) with sufficient accuracy, dn/dc should be higher than about 0.1 ml/g. This condition rules out the most common solvent of all nylons – cresol. Anhydrous acids – formic acid and sulphuric acid – cannot be employed because of the polyelectrolyte effect (See 5.1). Neither is it recommended to use other potential one-component solvents: chloroacetic acid because of the presence of aggregates in solution[27)], molten chloral hydrate because of the hydrolysis of nylons[48)], and, according to some authors 2,2,3,3-tetrafluoropropanal[5, 49)] owing to the polyelectrolyte effect.

Solvents of nylons for light scattering measurements are mostly multicomponent. An error in M_w, due to the optical effect of selective sorption, may be avoided by using the refractive index increment measured under the condition of dialysis equilibrium between the solution of nylon in a given mixed solvent and the mixed solvent[50)] (See Chapter 6).

The first attempts to measure the M_w of nylon 6,6 by light scattering in dilute formic acid were unsuccessful[51)], because of the presence of aggregates. Later on it was shown[52)] that nylon is aggregated in formic acid only at higher polymer concentrations and that below a concentration of about 5×10^{-3} g/ml it is dissolved molecularly. The polyelectrolyte effect was suppressed by water (15 wt.-% is sufficient) or by a low-molecular weight salt (KCl or HCOONa)[53, 54)] Addition of both water and salt decreases the second virial coefficient, A_2 (Figs. 2 and 3); A_2 assumes zero value corresponding to the θ-state for nylon 6 and 6,6 in 64 wt.-% HCOOH[55)] or, *e. g.* in 90 wt.-% HCOOH with 2.3M KCl[54)]. A detailed light scattering study of the stability of nylon 6 in formic acid solutions against hydrolysis showed that a more distinct decrease in M_w (above 10%) occurred at room temperature only after several days or on heating above 60 °C within several hours[13)].

The polyelectrolyte effect of nylon 6 in sulphuric acid can be suppressed with as little as 1 wt.-% water[10)]. A_2 decreases with increasing water content and becomes zero in about 33 wt.-% H_2SO_4[13, 56)]. The hydrolytic decrease in M_w increases with dilution of the acid. While in dilute H_2SO_4 M_w considerably diminishes within a few hours[13)], hydrolysis of nylon 6 in 96% H_2SO_4 is regarded as negligible (cf.[27)]). However, a disadvantage of concentrated H_2SO_4 as a solvent of nylon 6 for light scattering is the relatively low dn/dc (0.08 ml/g), as well the difficulties involved in the optical clarification of solutions[13)].

Fluorinated alcohols appeared to be suitable solvents of nylons in light scattering measurements, mainly due to very low refractive indexes of these sol-

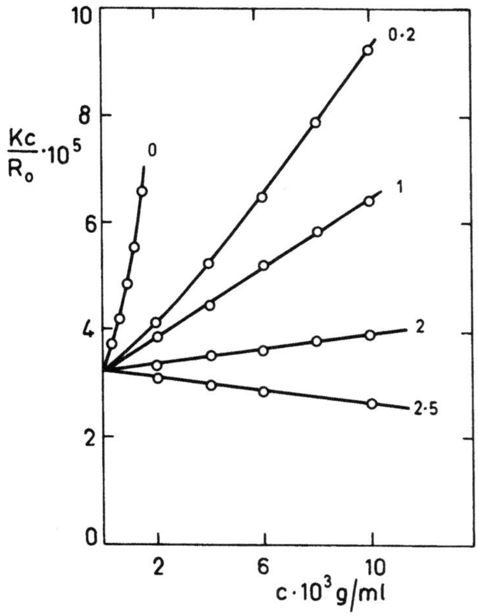

Fig. 2. Concentration dependence of light scattering of nylon 6,6 in the system 90 wt.-% HCOOH/KCl. Curves are labelled with KCl molarity [53)]

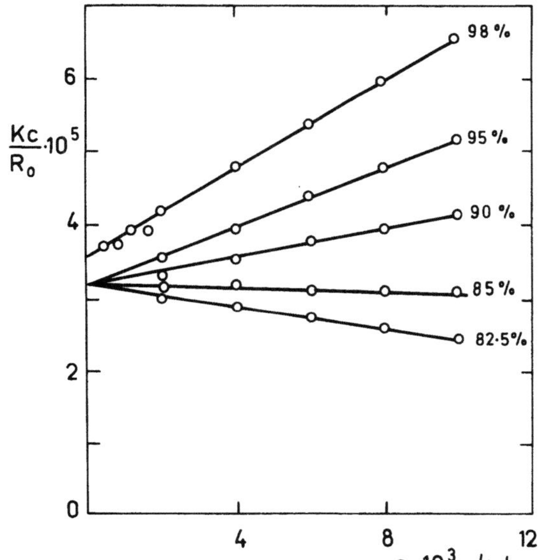

Fig. 3. Concentration dependence of light scattering of nylon 6,6 in the system HCOOH/2 M KCl. Curves are labelled with concentration of HCOOH in wt.-% [53)]

vents and thus high refractive index increments of nylons. Some authors performed their measurements in pure fluorinated alcohols, *e. g.* nylon 6,6 in 2,2,3,3-tetrafluoropropanol[57)] and in octafluoropentanol[27)], nylon 6 in 2,2,2-trifluoroethanol[58)] and nylon 12 in 1,1,1,3,3,3-hexafluoro-2-propanol[59)]; others, bearing in mind errors which might occur due to the high values of the second virial coefficient, the potential operativity of the polyelectrolyte effect and of partial aggregation, carried out light scattering measurements of nylon 6,6 in the mixture

2,2,3,3-tetrafluoropropanol/0.1M sodium trifluoroacetate [49], of nylon 6 in the ternary system 2,2,3,3-tetrafluoropropanol/0.1M LiCl/10 vol.-% H_2O [5], and of nylon 4 in the mixture 2,2,3,3-tetrafluoropropanol/50 vol.-% H_2O[4].

It was found that in many mixtures which are thermodynamically good solvents of nylon 12 (*e. g.*, 2,2,3,3-tetrafluoropropanol/benzene or 2,2,3,3-tetrafluoropropanol/toluene), the polymer undergoes spontaneous aggregation in a few hours[60]. Light scattering was performed with m-cresol/60 vol.-% 2,2,3,3-tetrafluoropropanol as solvent, which yields molecular solutions of nylon 12 stable in time[60]. Owing to the selective sorption of m-cresol on the polymer, the system possesses a rather high refractive index increment measured after the establishment of the dialysis equilibrium between solution and solvent; this makes possible the determination of molecular weights of an order of magnitude of a few thousands with good accuracy[60, 61].

Preliminary results show[62] that for a more accurate determination of M_w of nylon oligomers, systems with extremely high refractive index increments (>0.5 ml/g, *e. g.* in the mixture m-cresol/n-heptane) may prove helpful (See Chapter 6).

Since the M_w of common nylon samples virtually always varies within the order of magnitude of 10^4, the size of polymer molecules is too small to be determined by light scattering. The radius of gyration was measured[5] only for several fractions of nylon 6 with $M_w = 1 \times 10^5 - 3.5 \times 10^5$. The values thus obtained roughly coincided with those calculated from the Flory-Fox equation (*i. e.* from M_w, $[\eta]$ and the Flory universal constant).

4 Methods of Determination of Molecular Weight Distribution

A rough but often sufficient criterion of the width of molecular weight distribution is the ratio of two different molecular weight averages, most frequently of the weight- and number-averages (M_w/M_n). Each average may be determined either by indepedent methods, or their ratio may be found by a single method (See 4.2 and 4.3).

The molecular weight distribution curve can be obtained by a number of procedures. The methods applied to nylons have been (1) fractionation, (2) turbidimetric titration and (3) gel permeation chromatography.

4.1 Fractionation

The aim of preparative fractionation of nylons is to obtain fractions which may be used either in the determination of the molecular weight distribution curve or in the determination of the relationship between intrinsic viscosity and molecular weight. In both cases, the goal is to obtain as narrow fractions as possible; in the determination of the molecular weight distribution a full recovery of the sample is mandatory.

The determination of the distribution curve is also affected by the quality and number of fractions. The narrower the distribution of fractions the higher their quality and this depends on the chosen fractionation system and fractionation scheme.

A classical method of the preparative fractionation of polymers consists of fractional precipitation based on the separation of molecules with respect to their molecular weight between two liquid phases, namely, the more concentrated gel phase and the dilute sol phase. With nylons (similar to other crystalline polymers) the separation efficiency with respect to molecular weight is unfavorably affected in numerous systems by aggregation of polymer molecules near the precipitation point. The partly crystalline gel phase may then contain a certain amount of the coprecipitated lower-molecular weight fractions.

Several fractionation systems have been described for nylon 6 and 6,6; all of them may be applied to the other nylons, either directly or with small modifications. The oldest and still most frequently used system, phenol/H_2O, requires a working temperature of at least 70 °C owing to the limited miscibility of components at room temperature[12, 63, 64]. The temperature may be reduced if some "compatibilizing" component, such as ethylene glycol, is added to the system[65, 66]. An effort to avoid elevated temperature has led many authors to other systems, *e. g.* cresol/methanol[67], cresol/petroleum ether[68], cresol/cyclohexane[69] or (phenol/tetrachloromethane)/n-heptane[70]. While nylon tends to form a crystalline gel phase in the system cresol/methanol, the gel phase is more or less amorphous in the other systems mentioned here. Fractionation yielding an amorphous gel phase is sometimes called – rather misleadingly in our view – coacervate fractionation[65, 66]. An effective system seems to be cresol/n-heptane, in which on the threshold of phase separation nylon is demonstrably molecularly dissolved, *i. e.*, unaggregated[62]. Fractional extraction of the polymer from the gel phase, also called extraction from the coacervate, was applied to nylon 6 and to nylon 6,6[70–73]. In this procedure, the system is enriched with the solvent and fractions are collected continuously.

A view has been put forward that spontaneous aggregation at the onset of fractional precipitation, reducing the separation efficiency, is in some cases due to

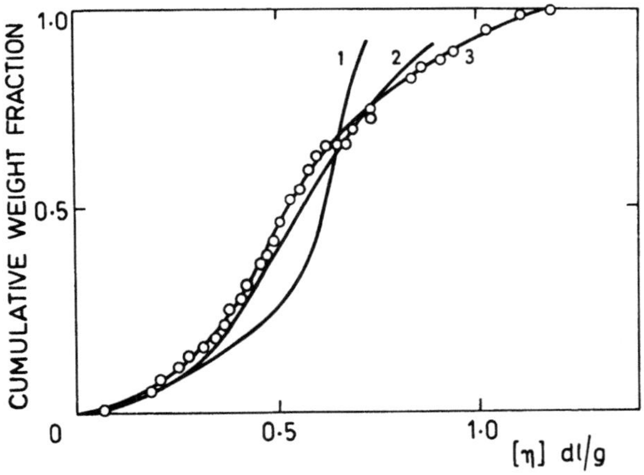

Fig. 4. Integral molecular weight distribution curves of nylon 6,6 on the basis of fractionation in m-cresol/cyclohexane. The number of fractions increases in the sequence 1, 2, 3[69]

the linear association of nylon containing both carboxyl and amine groups[74, 75]. The effect of concentration of acetylated end groups on the fractionation of nylon 6 was studied in the system (phenol/tetrachloromethane)/heptane. Even a small quantity of acetylated groups, which suppressed the association of nylon, considerably improved the fractionation efficiency. The authors therefore recommend introducing acetyl groups into samples intended for fractionation by an additional acetylation with acetylcaprolactam[74].

The effect of the number and quality of fractions was demonstrated in early attempts to determine the molecular weight distribution of equilibrium-polymerized nylons 6 and 6,6. While some authors[63, 76, 77] confirmed that equilibrium-polymerized nylons had the most probable distribution predicted by Flory[78], others[67, 79] found distributions which were considerably narrower. While the most probable distribution was established from a large number of fractions obtained by refractionation of primary fractions[63, 64, 76] narrower distributions resulted from fractionations with a smaller number of fractions[67, 79].

It is known from the theory of fractionation that, as a rule, the separation of very small fractions in successive fractionation does not yield the narrowest fractions[80]. A better result may be obtained by fractionating the sample into a smaller number of fractions and their subsequent refractionation. Howard[69] separated the same nylon 6,6 sample into different numbers of fractions. As the number of fractions increased the experimental integral distribution curves approached the shape of the most probable distribution (Fig. 4). Broadening of the experimental distribution function by refractionation of primary fractions has also been observed by other authors[76, 81].

One should bear in mind that not every nylon, even an equilibrium-polymerized one, must, possess the most probable molecular weight distribution. It was shown, for instance[27], that the distribution width given by the ratio M_w/M_n varied in nylon 66 samples from 1.5 to 2.3 according to the initial ratio adipic acid/hexamethylenediamine. Extremely wide distributions (M_w/M_n up to 6) were observed with nonequilibrium low-conversion nylons 12[82]. On the contrary, with the hydrolytic nylon 6 in one case a pronouncedly narrow distribution was reported (M_w/M_n close to unity)[83]; with respect to the hazardous application of the hot 90 wt.-% formic acid in fractional dissolution, when some hydrolysis of the sample is unavoidable, this finding is not persuasive.

4.2 Turbidimetric Titration

Turbidimetric titration is a non-preparative modification of the method of successive precipitation fractionation. The precipitation of a polymer from a highly dilute solution yields a turbidimetric curve, *i. e.* the dependence of the turbidity of solution on the amount of precipitant added. Recalculation of the turbidimetric curve into the molecular weight distribution curve is questionable[84] in view of some shaky assumptions viz., that the polymer is consecutively precipitated according to decreasing molecular weights and that turbidity is proportional to the amount of precipitated polymer. The latter requirement is not easily satisfied.

Unlike preparative fractionation, cumulative precipitation is a more or less non equilibrium process. The result is strongly affected by experimental conditions, such as the solvent/precipitant system, the rate of addition of the precipitant, intensity of stirring. Similar to preparative fractionation, it is also very desirable here that the polymer should not be phase-separated from solution in the crystalline form, but as a liquid amorphous phase which could be maintained in solution in the disperse state (usually by stirring).

Although simple solvent/precipitant systems such as m-cresol/cyclohexane[85] were also used for nylons, the effort to satisfy the demands of the method on the properties of the gel phase has led most authors to very complicated multicomponent systems, as *e. g.* (cresol/isobutanol/petroleum ether/cyclohexane)/(petroleum ether/1,1,2,2-tetrachloroethane)[86, 87]. The M_w/M_n values obtained from the turbidimetric titration data were lower than those from fractionation data[76, 85], resembling results obtained from fractionation with an insufficient number of fractions[69].

The data published so far on nylon 6 and 6,6 allow the conclusion that turbidimetric titration is not a suitable method for an accurate determination of the molecular weight distribution, but that owing to its speed, it may be employed, *e. g.*, in preliminary comparative studies of the relative changes of distributions[33].

4.3 Gel Permeation Chromatography

Gel permeation chromotography (GPC) is relatively novel and has been only recently applied to the study of nylons. After the necessary calibration and under certain asumptions, GPC data may provide the distribution curve of molecular weights and hence the molecular weight averages.

The choice of solvent seems to be the main restricting factor in the investigation of nylons by GPC. Requirements on the solvent are similar to those of light scattering (ability to dissolve nylon molecularly, a sufficiently high refractive index increment, etc.); moreover, the viscosity of the solvent should not be too high. It must not react with the column packing, being at the same time solvent for the polymer, the fractions of which are used in the calibration (polystyrene, as a rule). It should be pointed out in advance that a solvent for nylons meeting all these requirements has so far not been found.

For nylons 6 and 6,6, two single solvents have been used up to now, namely, m-cresol[88, 89] and 2,2,2-trifluoroethanol[46, 58]. The only advantage of m-cresol is that it is a solvent for polystyrene at the same time, and the calibration may therefore be performed directly. Its high refractive index and thus the low refractive index increment of nylon, on the other hand, called for the application of solutions with a higher nylon concentration, which brought about column overloading and distorted results[90]. Another disadvantage of m-cresol is seen in its high viscosity, which raises the pressure in the column and impairs the resolution power. At a working temperature above 100 °C viscosity was sufficiently reduced, and the results obtained with nylon 6[88] and 6,6[89] seemed to be consistent (*e. g.* the agreement between M_w from GPC and light scattering). On the other hand, a more than acceptable hydrolysis of nylon[47, 91] was proved to occur in cresol solutions above 100 °C. A reduction of the solution viscosity was achieved by mixing equal volumes of cresol and chlorobenzene. At the experimental temperature of 43 °C no degradation of nylon took place, and the results were satisfactory; the M_n obtained coincides with the value provided by end group analysis[91].

Unlike m-cresol, 2,2,2-trifluoroethanol has a very low refractive index (the refractive index increment of nylons is high) and a considerably lower viscosity. This allows the use of small polymer concentrations and considerably lower temperatures (50 °C[90], 25 °C[58]). Chromatograms appeared to react sensitively to differences in distributions, *e. g.* of the hydrolytic and alkaline nylon 6[58]. A serious complication in the determination of the molecular weight distribution curves is that 2,2,2-trifluoroethanol does not dissolve polystyrene. An attempt was made to employ a complicated calibration procedure, which yielded promising results[46]. On the other hand, it was pointed out[91] that 2,2,2-trifluoroethanol as solvent might cause wrong results owing to its incompatibility with the column packing (Styragel).

For nylon 11 and 12, hexamethylphosphortriamide[82, 92] was used as the solvent; due to its properties (advantageous refractive index, usability at room temperature), this solvent seems to be promising.

Although the investigation of nylons by GPC has only just begun, there is a hope that it may fully replace turbidimetric titration and partly replace preprarative fractionation in the analysis of molecular weight distributions of nylons.

5 Hydrodynamic Behavior

5.1 Intrinsic Viscosity

Owing to its experimental simplicity, viscometry seemed a suitable tool for a rapid and reliable characterization of nylon samples in research and particularly in practice. The intrinsic viscosity, $[\eta]$, is a measure of the hydrodynamic volume of the macromolecular coil and depends on the molecular weight of the polymer and on the thermodynamic quality of the solvent. $[\eta]$ value may be converted into the viscosity-average molecular weight, M_v, if the relation $[\eta] - M$ for the given polymer, solvent and temperature is known (See 5.2); the ratio of $[\eta]$ of linear and branched molecules with the same molecular weight allows the estimation of the degree of branching (See Chapter 7).

Intrinsic viscosity is determined by extrapolation of the concentration dependence of reduced quantities characterizing the solution viscosity to zero polymer concentration. The classical Huggins[93] and Kraemer[94] equations are those most frequently applied:

$$\frac{\eta_{sp}}{c} = [\eta] + k' [\eta]^2 c + \dots \text{ and} \tag{1}$$

$$\frac{\ln \eta_{rel}}{c} = [\eta] - k'' [\eta]^2 c + \dots \tag{2}$$

where η_{sp} is the specific viscosity, η_{rel} is the relative viscosity, c is the polymer concentration (traditionally given in g/dl), and k' is the Huggins coefficient, related with the parameter k'' by $k' + k'' = 0.5$.

Application of other expressions reviewed in Ref.[95] is dictated by an effort to carry out the most reliable and linear extrapolation possible, taking into account also higher terms of the power series expansions of the respective equations. Recently, various extrapolation procedures have been discussed in detail in connection with the measurements of relatively low-molecular weight samples of nylon 12, in which the shape of the function $\eta(c)$ in cresol depended not only on $[\eta]$, but also on the character of end groups (*i. e.* on the initiation system and polymerization mechanism[95]). With acidolytic polymers, the most reliable results were obtained from a plot recorded by Heller[96] ($1/2(c/\eta_{sp} + c/\ln \eta_{rel})$ vs. c). For hydrolytic nylons with $[\eta] > 0.8$ the most reliable plot was that from Martin, Staudinger and Heuer[97, 98] ($\ln \eta_{sp}/c$ vs. c), along with the plot from Heller, while at $[\eta] < 0.8$, where the concentration dependences of viscosity are anomalous, only the Martin-Staudinger-Heuer plot[97, 98] could be employed.

The determination of $[\eta]$ from measurements of nylon solutions at a single concentration is also found in the literature. Of course, the accuracy of such determination, which depends of the validity of certain assumptions, is much lower than the determination by extrapolation methods. Thus, some authors[1, 27] calculate $[\eta]$ from measurements at a single concentration using the equation

$$[\eta] = (2/c)\,(\ln \eta_{rel} - \eta_{sp})^{1/2} \tag{3}$$

Since this relation has been derived from Eqs. (1) and (2) assuming that higher terms of the two polynomials are negligible, its applicability is restricted[99] to solutions with $\eta_{sp} < 0.6$ and $k' \leqslant 0.4$. Other authors[100] evaluate $[\eta]$ from η_{sp}/c measured at a single concentration by means of Eq. (1); here, k' is substituted by an average value experimentally determined for a given system in a given range of molecular weights. With respect to the dependence of the Huggins coefficient on many properties of nylon (*e. g.* molecular weight, branching, in some cases on the properties of end groups, etc.), such determination of $[\eta]$ may be subject to considerable error and cannot be recommended for general use[101]. One-point estimates of $[\eta]$ can especially with low-molecular weight nylons lead to large errors.

The values of the Huggins coefficient k' determined in cresol for low-molecular weight samples of nylon 6[102] and nylon 12[61], prepared by hydrolytic polymerization of lactams, were extremely high. The phenomenon has been interpreted as association of carboxyl and amine end groups. The association constants of the two nylons were calculated from the k' values[95, 102].

Compared with other methods, the requirements for solvents for the determination of $[\eta]$ are not so specific. Virtually all solvents used in the viscometry of nylons are given in Table 2, Section 5.2. A solvent for nylons by tradition most widely used is m-cresol. Even in this solvent, however, the values may be subjected to errors. m-Cresol is only seldom used as a chemical individuum. Measurements are usually carried out in technical m-cresol, which is a mixture of o-, m- and p-isomers with about 60% of the m-component, or in the so-called tricresol, which is a

Table 2. Constants of Mark-Houwink equation (K, a) constant K_0 and characteristic ratio C_∞ [eq. (4)]

Solvent	Temp °C	Method	Range of $M \times 10^{-3}$	Samples	$K \times 10^4$ (dl/g)	a	$K_o \times 10^4$ (dl/g)	C_∞	Ref.
Nylon 4									
m-Cresol	25	M_w–LS	10–300	f,u	4.0	0.77	22.8	7.2	4)
3 g $CaCl_2$/10 ml (H_2O/35 vol.-% C_2H_5OH)	25	M_w–LS	9–155	f,u	10.7	0.57	19.6	6.5	122)
TFP/10 vol.-% H_2O	25	M_w–LS	10–300	f,u	1.8	0.78	14.3	5.2	4)
m-Cresol	25	M_n–EG	1.7– 14	u	30.0	0.70			147)
Nylon 5									
m-Cresol	25	M_n–EG	6– 46	u	44.8	0.59			148)
Nylon 6									
m-Cresol	25	M_w–S	6– 60	f	1.0	1.03			110)
m-Cresol	25	M_w–LS	5– 40	u	5.57	0.73			23)
m-Cresol	25	M_w–LS	9–355	f	5.26	0.74	26	7.2	5)
m-Cresol	25	M_w–S	3– 60	f	5.8	0.73	21	6.3	109)
m-Cresol/25 vol.-% Decalin	25	M_w–LS	30–200	f	29.4	0.52	31	8.1	121)
Phenol/Tetrachloroethane (1/3 by wt.)	25	M_w–LS	30–200	f	7.2	0.72	25	7.0	121)
HFP/0.05M NaTFAc	25	M_w–S	10– 30	f	48.0	0.65			110)
85 wt.-% HCOOH	25	M_w–LS	16–300	f	4.7	0.73	20	6.1	121)
TFE	25	M_w–S	3– 60	f	34.0	0.56	35	8.8	109)

$TFE/CHCl_3$ (3/7 by vol.)	25	M_w–S	3– 60	f	53.0	0.52	45	10.5	109)
TFP/0.1M LiCl/ 10 vol.-% H_2O	25	M_w–LS	9–355	f	3.42	0.76	20	6.1	5)
m-Cresol	30	M_n–EG	4– 25	u	10.7	0.70			10)
m-Cresol	25	M_n–EG	2– 27	u	2.2	0.82			30)
m-Cresol	30	M_n–EG	0.3– 4	f	0.3	0.53			108)
m-Cresol	25	M_n–EG	5– 25	u	31.9	0.62			25)
m-Cresol	25	M_n–calculated			24.0	0.62			25)
m-Cresol	25	M_n–EG	5– 30	u	18.0	0.65			23)
m-Cresol	25	M_n–EG	12– 20	u	12.6	0.70			112)
Tricresol	25	M_n–Os	8– 80	u	2.1	0.90			107)
m-Cresol	25	M_n–EG	2– 20	u	7.6	0.75			102)
Chloralhydrate/ 30 vol.-% H_2O	25	M_n–EG	20– 30	u	5.5	0.76			149)
85 wt.-% HCOOH	25	M_n–EG			0.6	0.98			56)
40 wt.-% H_2SO_4	25	M_n–EG	3– 13	f	5.9	0.69			150)
Conc. H_2SO_4	18	M_n–EG	0.4– 5	u	1.2	0.67			8)
Conc. H_2SO_4	25	M_n–EG	4– 37	u	6.3	0.76			113)
Conc. H_2SO_4	30	M_n–EG	4– 25	u	5.2	0.77			110)
Conc. H_2SO_4	25	M_n–EG	2– 17	f	5.0	0.79			74)
Conc. H_2SO_4	25	M_n–Os	8– 80	u	0.9	0.98			107)
Conc. H_2SO_4	25	M_n–EG	8– 18	u	5.1	0.79			151)
98 wt.-% H_2SO_4	25	M_n–EG	11– 22	u	1.0	1.0			152)
TFE	25	M_n–EG	1– 4	u	0.97	1.0			153)
m-Cresol	25	M_v–V	5– 25	f	32.0	0.62			79)
15 g $CaCl_2$/dlCH_3OH	35	M_v–V	13– 27	f	16.8	0.50	36.8	9.14	123)
20 g $CaCl_2$/dlCH_3OH	35	M_v–V	11– 54	f	38.6	0.51	36.8	9.14	123)
64 wt.-% HCOOH	25	M_v–V	16– 56	u	20.3	0.50			55)
85 wt.-% HCOOH	20	M_v–V	4– 16	u	7.5	0.70			14)
85 wt.-% HCOOH	20	M_v–V	7–120	f	2.26	0.82	23.0	6.7	72)
85 wt.-% HCOOH	–10	M_v–V	7–120	f	2.68	0.82	23.0	6.7	126)
85 wt.-% HCOOH	0	M_v–V	7–120	f	2.48	0.82	23.0	6.7	126)
85 wt.-% HCOOH	10	M_v–V	7–120	f	2.34	0.82	23.0	6.7	126)

Table 2 (continued)

Solvent	Temp °C	Method	Range of $M \times 10^{-3}$	Sam-ples	$K \times 10^4$ (dl/g)	a	$K_o \times 10^4$ (dl/g)	C_∞	Ref.
Conc. H_2SO_4	30	M_v–V	4– 10	f	1.3	0.91			12)
Conc. H_2SO_4	40	M_v–V	4– 10	f	3.3	0.80			12)
Conc. H_2SO_4	50	M_v–V	4– 10	f	6.3	0.72			12)
TFE	–20	M_v–V	13–100	f	5.33	0.74	50.0	11.2	126)
TFE	25	M_v–V	13–100	f	5.36	0.75	34.0	8.6	126)
TFE	50	M_v–V	13–100	f	5.82	0.73	34.0	8.6	126)
Nylon 6, cyclic oligomers									
m-Cresol	25	M_n–Os	0.22–0.56		210	0.22			154)
Ethylene Chlorohydrin	25	M_n–Os	0.22–0.56		87	0.27			154)
Nylon 6,6									
o-Chlorophenol	25	M_w–LS	7– 80	f, u	16.8	0.62			1)
m-Cresol	25	M_w–LS	0.15– 50	u	$[\eta] = 1.5 \times 10^{-2} + 3.53 \times 10^{-4} M^{0.79}$		28.5	7.7	27)
m-Cresol	25	M_w–LS	7– 80	f, u	24	0.61			1)
90 wt.-% HCOOH	25	M_w–LS	7– 68	f, u	3.53	0.79			1)
90 wt.-% HCOOH	25	M_w–LS	0.15– 50	u	$[\eta] = 2.5 \times 10^{-2} + 2.5 \times 10^{-4} M^{0.87}$		19.7	6.0	27)
90 wt.-% HCOOH/ 0.1M HCOONa	25	M_w–LS	7– 80	f, u	8.8	0.65			1)
90 wt.-% HCOOH/ 0.1M HCOONa	25	M_w–LS	0.15– 50	u	$[\eta] = 1 \times 10^{-2} + 5.16 \times 10^{-4} M^{0.69}$				27)
90 wt.-% HCOOH/ 2M KCl	25	M_w–LS	2.5– 50	u	14.2	0.56	23.1	6.7	27)
90 wt.-% HCOOH/									

2.3M KCl	25	M_w–LS	2.5– 20	u	22.2	0.50	22.2	6.5	27)
90 wt.-% HCOOH/ 2.3M KCl	25	M_w–LS	7– 80	f, u	22.7	0.50	22.7	6.6	1)
90 wt.-% HCOOH/ 2.3M KCl	25	M_w–LS	20– 83	f	19.2	0.50	19.2	5.9	114)
Conc. H_2SO_4	25	M_w–LS	7– 80	f	11.5	0.67			1)
Conc. H_2SO_4	25	M_w–LS	0.15– 50	u	$[\eta] = 2.5 \times 10^{-2} + 2.5 \times 10^{-4} M^{0.83}$		29.0	7.8	27)
TFP/0.1M Na TFAc	25	M_w–LS	7– 80	f	11.4	0.66			1)
90 wt.-% HCOOH/ 0.1M HCOONa	25	M_n–EG	10– 50	u	3.22	0.74			155)
90 wt.-% HCOOH	25	M_n–EG	6– 24	u	11.0	0.72			45)
m-Cresol	20	M_v–V	10– 40	f, u	38.0	0.55			124)
85 wt.-% HCOOH	25	M_v–V	10– 40	f, u	6.1	0.71			124)
95 wt.-% HCOOH	25	M_v–V	10– 40	f, u	6.7	0.72			124)
61 wt.-% H_2SO_4	40	M_v–V	10– 40	f, u	18.3	0.55			124)
84 wt.-% H_2SO_4	40	M_v–V	10– 40	f, u	21.3	0.57			124)
Conc. H_2SO_4	40	M_v–V	10– 40	f, u	22.0	0.60			124)
Conc. H_2SO_4	20	M_v–V	10– 40	f, u	23	0.65			124)
Polycond. nylons 6 + 6,6									
m-Cresol	20	M_n–Cr, Os	1– 36	f	10.6	0.71			39)
Nylon 6,10									
m-Cresol	25	M_n–EG	8– 24	u	1.35	0.96			144)
Nylon 8									
m-Cresol	40	M_n–EG	10– 55	f	0.81	0.94			156)
m-Cresol	25	M_n–EG	1– 25	u	7.0	0.76			157)

Table 2 (continued)

Solvent	Temp °C	Method	Range of $M \times 10^{-3}$	Sam-ples	$K \times 10^4$ (dl/g)	a	$K_o \times 10^4$ (dl/g)	C_∞	Ref.
Nylon 12									
m-Cresol	25	M_w–LS	10–125	f, u	4.6	0.75	23.8	6.3	60)
m-Cresol	25	M_w–LS	3–125	f, u	8.1	0.70	26.6	6.8	61)
m-Cresol	25	M_w–LS	15–160	f	6.2	0.73	42.0	9.2	59)
HFP	25	M_w–LS	15–160	f	8.4	0.68	39.0	8.7	59)
Conc. H_2SO_4	25	M_w–LS	10–125	f, u	6.9	0.64	22.0	6.0	60)
m-Cresol	25	M_n–EG	1– 33	u	11.8	0.73			61)
m-Cresol	20	M_n–EG	5– 17	u	15.5	0.70			158)
Phenol/tetrachlorethane (1/1 by wt.)	20	M_n–EG	5– 17	u	2.2	0.91			158)
Conc. H_2SO_4	20	M_n–EG	5– 17	u	2.9	0.80			158)

LS light scattering, S sedimentation, EG end group analysis, Cr cryoscopy, Os osmometry, V [η]-M relation, f fractions, u unfractionated samples, HFP 1,1,1,3,3,3 hexafluoro-2-propanol, TFE 2,2,2 trifluoroethanol, TFP 2,2,3,3 tetrafluoropropanol, NaTFAc sodium trifluoroacetate, conc. H_2SO_4 covers 95 wt.-% – 97 wt.-% H_2SO_4.

mixture of isomers containing about 40% of the m-component and a varying amount of other phenolic compounds. It was shown that $[\eta]$ of the same polymer in cresols with various composition might differ considerably[103].

In some solvents, *e. g.* in anhydrous formic and sulphuric acid, nylon solutions exhibit anomalous concentration dependences of reduced viscosity parameters owing to the polyelectrolyte effect. In an acid medium capable of dissociation the amide group may add a proton

$$-CONH- + AH \rightleftharpoons -CONH_2^+- + A^-$$

In such a case, positively charged groups along the nylon chain repel each other, thus expanding the polymer coil. The concentration of counterions A^- which

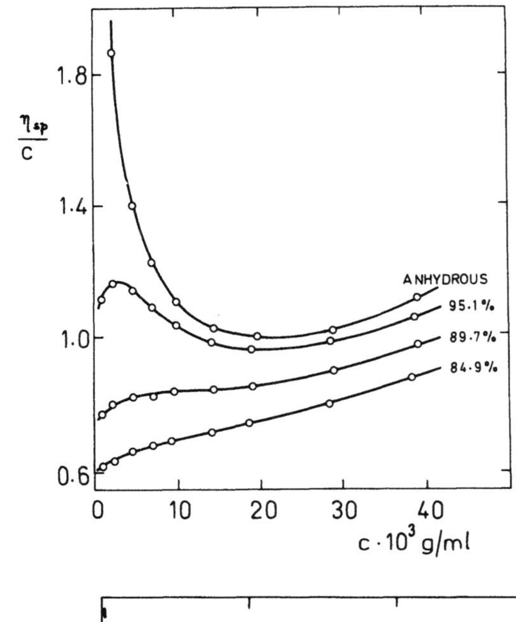

Fig. 5. Concentration dependence of reduced specific viscosity of nylon 6 in HCOOH. Curves are labelled with concentration of HCOOH in wt.-%[10]

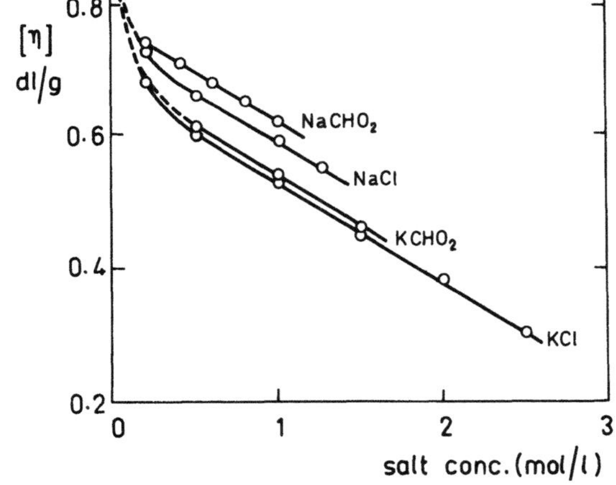

Fig. 6 Dependence of intrinsic viscosity of nylon 6,6 in 90 wt.-% HCOOH on the concentration of different salts in solution[53]

partly screen the electrostatic repulsion of $-CONH_2^+$-groups, diminishes with decreasing nylon concentration in solution; the lower the concentration of nylon, the more expanded the polymer coils are. This so-called polyelectrolyte effect manifests itself markedly in an anomalous concentration dependence of reduced specific viscosity[10, 104] (Fig. 5) and renders difficult or even impossible the determination not only of specific viscosity, but also of molecular weight and dimensions from light scattering data. The polyelectrolyte effect may be suppressed by rasing the ionic strength of solution. In solutions of nylon 6 and 6,6 in formic acid it was suppressed by water[10, 51] (Fig. 5), or – more effectively – by addition of a low-molecular weight salt[10, 51] (Fig. 6). A sharp drop in $[\eta]$ values in Fig. 6 within the salt concentration range 0–0.4 mol/l may be explained by screening of charges on the polymer by salt anions; an additional, not so sharp drop may be due to the deterioration of the thermodynamic quality of the solvent.

While some authors[10] assumed and observed in anhydrous formic acid a continuous increase in reduced specific viscosity with decreasing nylon concentration (Fig. 5), others[55] under the same conditions found a sharp maximum on the same dependence in the range of very low concentrations. An unpronounced maximum observed on the concentration dependence of η_{sp}/c in a solution of nylon 6 in anhydrous sulphuric acid indicates an only incomplete suppression of the polyelectrolyte effect due to partial dissociation of the acid itself[10]. The effect can be totally suppressed by a mere 1 wt.-% of water.

The literature dealing with the polyelectrolyte behavior of nylons in fluorinated alcohols is somewhat controversial. In solutions of nylon 6 and 6,6 in 2,2,3,3-tetrafluoropropanol, some authors observed the polyelectrolyte effect which they suppressed with trifluoroacetate[49], lithium chloride[5], or water[5]. Other authors[57] explained the polyelectrolyte effect in 2,2,3,3-tetrafluoropropanol solutions by the presence of traces of water while assuming that salt suppresses this effect not by raising the ionic strength of solution, but by binding traces of water. The polyelectrolyte effect was not observed in 2,2,3,3-tetrafluoropropanol solutions of nylon 4[4], 2,2,2-trifluoroethanol solutions of nylon 6[58], octafluoropropanol solutions of nylon 6,6[27] and 1,1,1,3,3,3-hexafluoro-2-propanol solutions of nylon 12[59].

5.2 Relationship Between Intrinsic Viscosity and Molecular Weight

For molecules of one type (*e. g.* linear) of a given polymer, intrinsic viscosity and molecular weight are related by the Mark-Houwink equation

$$[\eta] = KM^a \quad (4)$$

K and *a* are usually constants within a wide range of molecular weights, which at a given temperature depend only on the polymer/solvent system. From a measured $[\eta]$ value, it is possible with known *K* and *a*, to determine the viscosity-average molecular weight M_v; for a polydisperse sample, M_v lies between the number- and weight-average molecular weight, rather close to the weight-average. Since the

$[\eta]$-M_w relation is much less sensitive to polydispersity than the relation $[\eta]$-M_n (as shown in Figs. 8–11), methods yielding M_w should be preferred for calibration.

The constants of the Mark-Houwink equation published so far for nylons 4, 6, 6,6, 8, 6,10 and 12 are summarized in Table 2. Unlike most other polymers, *e. g.* vinyl polymers[105)], calibration by means of M_n is more frequent with nylons, obviously because the determination of M_w (by light scattering) is connected with a number of complicating factors (See 3.3), while the determination of M_n of nylons by end group analysis is a traditional, technically not too demanding method, which has been in use and described in the literature for some 40 years now.

With flexible macromolecules, the exponent *a* reflects the intensity of polymer-solvent interaction, *i. e.* the thermodynamic quality of the solvent for a given polymer. For nylons, as rather flexible chains, (See 5.3), *a* within the range of about 0.5–0.8 may be expected; published higher exponents, insofar as they were obtained in nonionizing solvents, cannot be regarded as too likely. The scattering of *a* values for identical systems, *e. g.* nylon 6 in m-cresol at 25 °C (0.53–1.03) or nylon 6 in concentrated sulphuric acid at 25 °C (0.67–1.0) shows that it is not advisable to estimate the thermodynamic quality of the solvent from the experimental values of *a*.

There are many reasons which may lead to different or erroneous results in the viscometry and $[\eta]$-M relations for nylons. Before illustrating the most serious of these reasons with practical examples (Figs. 8–13), let them be reviewed in brief below.

(1) The nature of the polymer. Depending on the polymerization conditions, nylon molecules may be terminated by various groups. The varying affinity of the solvent to these groups compared with monomeric units inside the chain, as well as their interaction (in some cases even association[102)]) may be reflected in the range of very low molecular weights mainly in the value of *a*. A pronounced effect on the $[\eta]$-M relation is exerted by branching[72, 106)].

(2) Determination of $[\eta]$. The possibility of an erroneous determination of $[\eta]$ in the commonest solvent, m-cresol, which is often replaced by mixtures of o-, m- and p-isomers or of other phenolic compounds has been pointed out in Section 5.1. In sulphuric and formic acid, there is the danger of hydrolysis, especially with high-molecular weight samples and in the preparation of solutions at elevated temperature[13)]. Considerable errors may arise, if $[\eta]$ is calculated from measurements at a single polymer concentration[30, 107)], especially in those cases in which the constant character of the Huggins coefficient *k'* cannot be guaranteed (See 5.1).

(3) Molecular weight determination. In the calibration of the $[\eta]$-M equation, end group analysis and light scattering are the most frequently used methods. The sources of errors in the determination of M_n and M_w by these methods have been discussed in Sections 3.1 and 3.3. The hazard of cumulation of errors arises from the indirect determination of molecular weight, *i. e.* M_v, from intrinsic viscosity by means of other $[\eta]$-M relations, derived moreover for a material of possibly different polydispersity. A frequent source of differing results is the narrow range of molecular weights employed in calibration.

(4) Polydispersity of samples. The common polydispersity of samples does not affect the relation $[\eta]$-M_w within the limits of experimental error. It has been con-

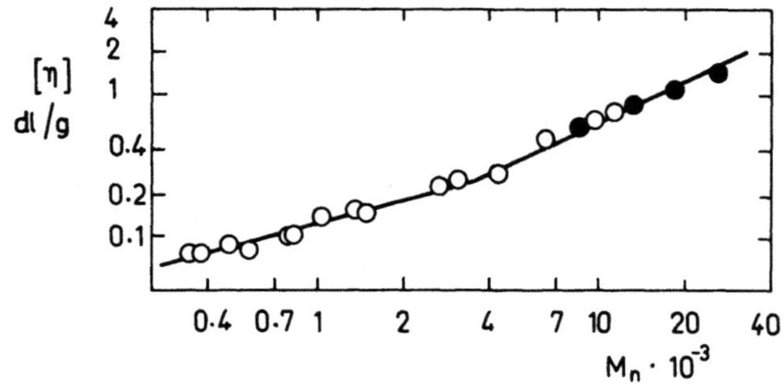

Fig. 7. $[\eta]$-M_n relationship for nylon 6 in m-cresol at 30 °C. Data from Ref.[108] ○, and Ref.[10] ●

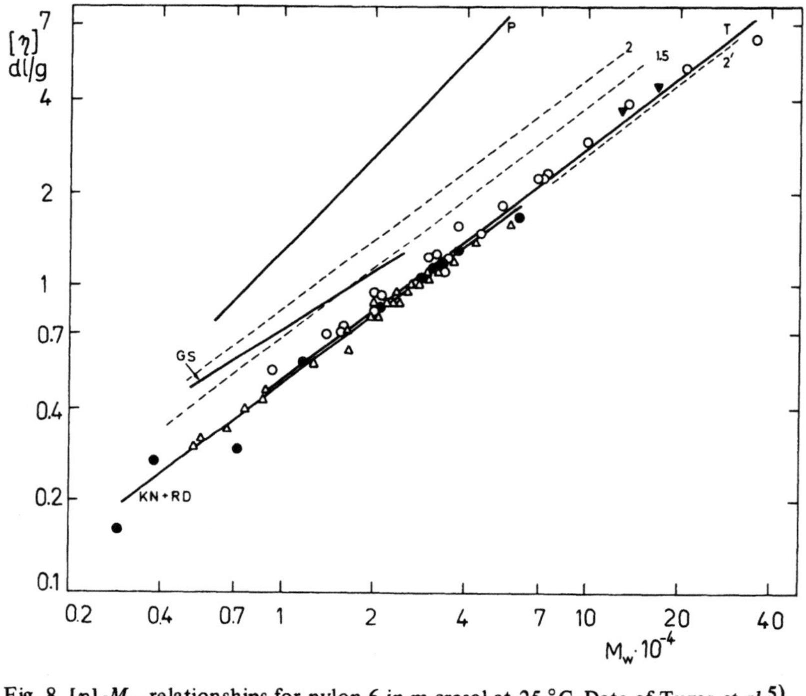

Fig. 8. $[\eta]$-M_w relationships for nylon 6 in m-cresol at 25 °C. Data of Tuzar *et al.*[5] —○— T; Kulakova, Nekrasov[109] —●— KN; Reimschuessel, Dege[23] —△— RD; Goebel *et al.*[107] ▼; Pavlov *et al.*[110] P; Griehl, Sieber[25] GS. Lines 1.5, 2 (for $[\eta]$-M_n) and 2' (for $[\eta]$-M_w) are calculated for M_w/M_n 1.5 and 2. M_w/M_n of relation from Ref.[5] is assumed 1

firmed experimentally that the given $[\eta]$-M_w relation is fulfilled both by fractions and by unfractionated polymers (*e. g.,* [5,60]). In the calibration with number-average molecular weights, however, polydispersity combined with the narrow range of molecular weights may be the largest source of errors. It is virtually impossible to prepare fractions with approximately equal polydispersities by employing usual fractionation procedures; and no attempts to characterize the range of molecular weight distribution of fractions for the calibration of the $[\eta]$-M_n relation have been reported. This uncertainty may explain the large scattering of data (*e. g.* on nylon 6 in m-cresol) not only for *K*, but also for *a*. It seems that calibration of the $[\eta]$-M_n relation is more reliable with unfractionated samples, for which distributions of approximately the same width may be assumed[61].

In two studies[27, 108] a curvature was found on the $[\eta]$-M dependence for samples with molecular weight of the order of magnitude 10^3. In the first paper[27], the curvature is based on the $[\eta]$ values of nylon 6,6 determined from measurements at a single concentration and the dependences were linearized by a plot in the form $[\eta] = A + KM^a$; the values of the constant *A* diminished with deteriorating quality of the solvent. In the second case[108], for nylon 6 in m-cresol, *a* changed from 0.70 to 0.53 in the range $M_n < 4\times10^3$ (Fig. 7). With the exception of the above two papers, such a curvature of the $[\eta]$-M dependences has not been observed.

It is difficult to estimate to what extent the respective relations in Table 2 are correct, and thus recommendable. The only virtually possible criterion of plausibility was seen in the degree to which results from independent papers agree with each other. Figs. 8, 9, 12 and 13 compile independent data on $[\eta]$-M_w dependences for nylon 6 and 6,6 in common solvents, such as m-cresol, sulphuric and formic acid. Figures 10 and 11 illustrate the large effect of polydispersity on scattering in the $[\eta]$-M_n relations for nylon 6.

The $[\eta]$-M_w dependences for solutions of nylon 6 in m-cresol (Fig. 8) have been taken from five papers[5, 23, 25, 109, 110]. Molecular weights were determined by light scattering[5, 23], by the Archibald method of approach to sedimentation equilibrium[109] and by means of the Mandelkern-Flory invariant from the sedimentation coefficient and intrinsic viscosity[110]. In another work[25], M_w was substituted by M_n determined by end group analysis assuming negligible polydispersities of the measured fractions. The data of three papers agree well with each other, while those of another paper[110] differ in M_w is as much as an order of magnitude. In the relationship just mentioned[110], which is too high the *a* seems questionable. Departures of data in Ref.[25] may be readily explained by the fact that the polydispersity index M_w/M_n of fractions taken for measurements was actually much higher than unity. The relations $[\eta]$-M_n and $[\eta]$-M_w calculated for various polydispersities[111] were related to one[5] of the three virtually identical relations (Figs. 8–10) obtained from the largest number of points within the broadest possible range of M_w. The $[\eta]$-M_w data on nylon 6,6 in m-cresol of the only two authors[1, 27] (Fig. 9) agree satisfactorily with the relation chosen for nylon 6[5]. This may be regarded as one of the proofs that the solution behavior of nylon 6 and 6,6 is very similar.

While the common polydispersity is not reflected in the $[\eta]$-M_w relations (Fig. 8), it probably affects the $[\eta]$-M_n relationships in a decisive way. It can be seen from Fig. 10, which gives the $[\eta]$-M_n relations for nylon 6 in m-cresol[10, 23, 25, 30, 79, 102, 107, 112], that virtually all data lie above dependence calculated for $M_w/M_n = 1.5$; some are even situated in the range $2 < M_w/M_n < 3$. It should also be taken into consideration that, along with polydispersity, the small range of molecular weights and errors ensuing from the determination of M_n by the end group method may also affect the scattering of data. The high polydispersity of samples from Ref.[107], suggested by the position of experimental points in Fig. 10, is corroborated by GPC data which give M_w/M_n predominantly between 2-2.5, in

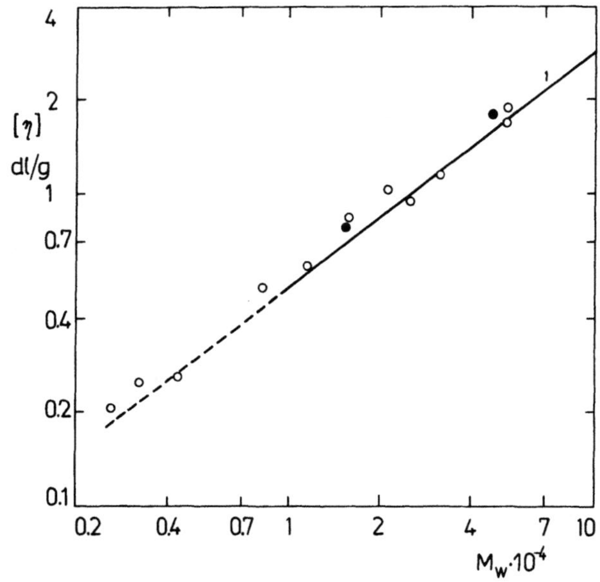

Fig. 9. [η] -M_w relationships for nylon 6,6 in m-cresol at 25 °C. Data of Elias, Schumacher[27] ○; Burke, Orofino[1] •; Tuzar *et al.*[5] for nylon 6 see line 1

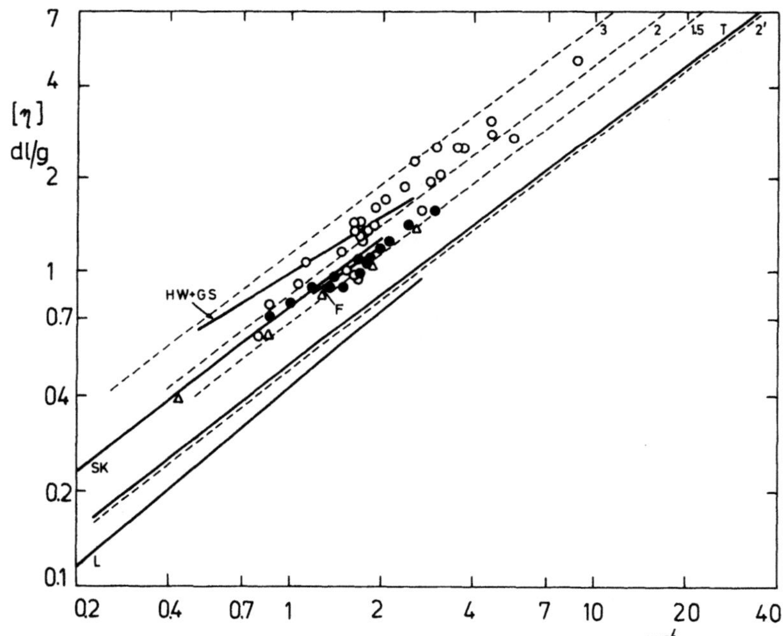

Fig. 10. [η] -M_n relationships for nylon 6 in m-cresol at 25 °C. Data of Goebel *et al.*[107] ○; Reimschuessel, Dege[23] •; Schaefgen, Trivisinno[10] △; Hoshino, Watanabe[79] HW; Šebenda, Králíček[102] SK; Loepelman[30] L; Griehl, Sieber[25] GS; Fukumoto[112] F. For lines 1, 1.5, 2, 3 and 2' see Figs. 8 and 9

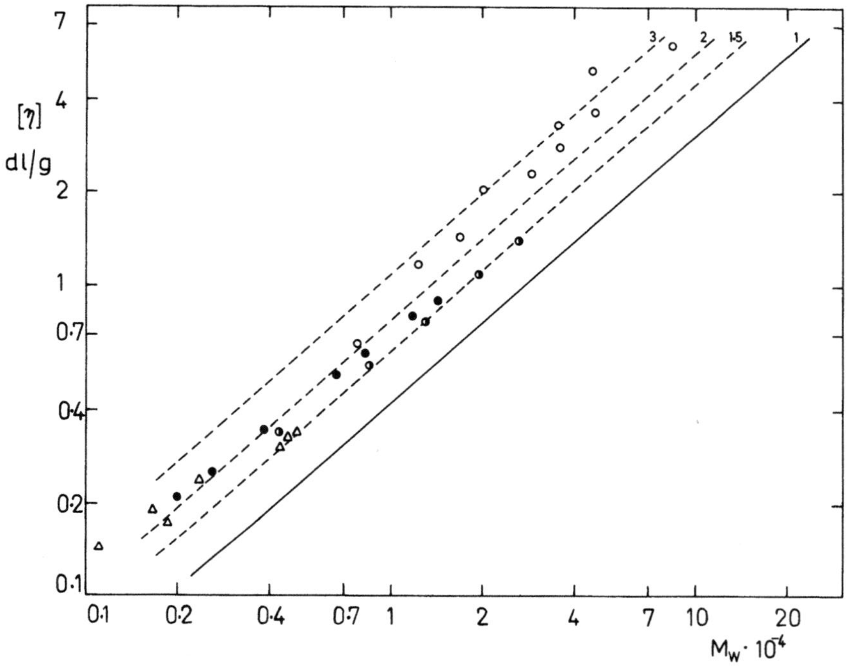

Fig. 11. $[\eta]$-M_n relationships for nylon 6 in conc. H_2SO_4. Data of Goebel *et al.*[107] ○; Schaefgen, Flory[113] (25 °C) ●; Schaefgen, Trivisonno[10] (30 °) ◑; Matthes[8] (18 °C) △. For lines 1.5, 2 and 3 see Fig. 8, for line 1 see text

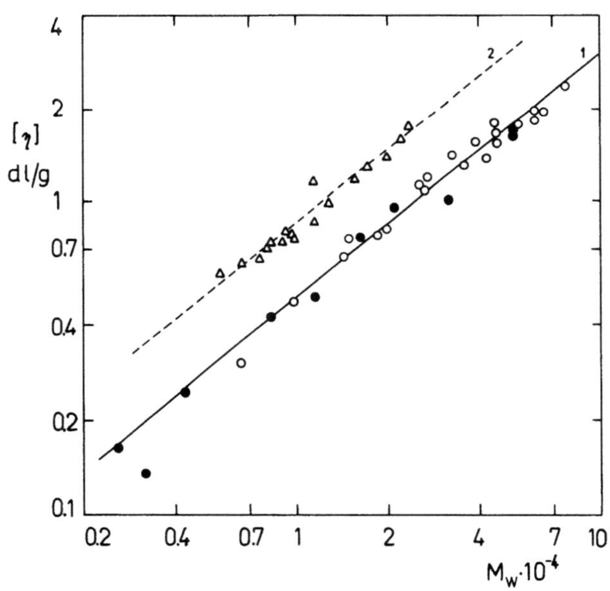

Fig. 12. $[\eta]$-M_w relationships for nylon 6,6 in 90 wt.-% HCOOH at 25 °C. Data of Burke, Orifono[1] ○; Elias, Schumacher[27] ● and $[\eta]$-M_n data of Taylor[45] △

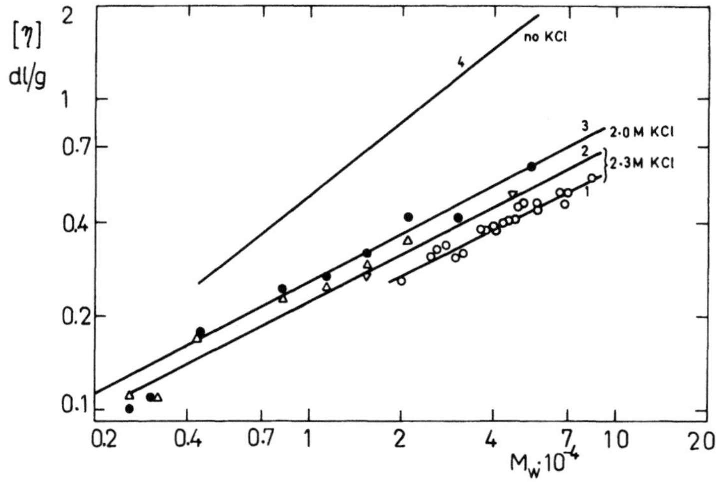

Fig. 13. $[\eta]$-M_w relationships for nylon 6,6 in 90 wt.-% HCOOH with KCl at 25 °C. Data of Saunders[114] ○ 1, Burke, Orofino[1] ▽ 2, Elias, Schumacher[27] △ 2, ● 3 and 4

some cases up to 2.9. Only the dependence in Ref.[30] remains unexplained; it is very close to the $[\eta]$-M_w relationship, although it was obtained with unfractionated samples by means of the end group method.

Concentrated sulphuric acid was a solvent frequently employed especially in earlier studies. All $[\eta]$-M relationships for nylon 6 in concentrated sulphuric acid available in the literature are based on M_n. The relationship $[\eta]$-M_w is not known; its approximate plot (straight line 1, Fig. 11) was constructed as follows: from data in Ref.[10], the relationship between intrinsic viscosities of nylon 6 in m-cresol and sulphuric acid was obtained. This was further used in transforming the relationship $[\eta]$-M_w in m-cresol into the $[\eta]$-M_w relationship in concentrated sulphuric acid. If experimental dependences $[\eta]$-M_n[8, 10, 107, 113] are compared with the approximate relation $[\eta]$-M_w (Fig. 11), then, in spite of different temperatures at which the data were obtained, the image and the conclusions regarding the effect of polydispersity are similar to those for solutions in m-cresol.

The $[\eta]$-M_w relation of nylon 6,6 in 90 wt.-% formic acid is documented by data of two papers[1, 27]. If one omits three values for oligomers[27], the intrinsic viscosity of which was determined from measurement at a single concentration, the data of the two papers virtually satisfy the same dependence (Fig. 12). Experimental points of the relation $[\eta]$-M_n for unfractionated samples[45] (also in Fig. 12) have approximately the same slope and shift as expected by a value corresponding to the ratio $M_w/M_n = 2$.

For nylon 6,6 in 90 wt.-% formic acid with 2.3M KCl (θ-solvent), the agreement between the $[\eta]$-M_w relationships, of various origin is somewhat poorer (Fig. 13). Burke and Orofino[1] recommend correcting the M_w values from Ref.[114] by a factor of 0.95, with respect to differences in the calibration of the lights scattering instruments; this correction reduces the difference to some extent, but does not remove it completely. The reasons of such disagreement may only be guessed. One of the likely causes may be, *e. g.* not exactly the same composition of the mixed solvent, because $[\eta]$ is very sensitive to the concentration of KCl[53]. This reasoning is also corroborated by the different dependence for 90 wt.-% formic acid with 2M KCl in Fig. 13.

Data for less common solvents (*e. g.* fluorinated alcohols) or solvents in which the hydrolysis of nylons has been proved to occur (*e. g.* dilute sulphuric acid) are not commented on here.

The analyses just outlined show that the most reliable relationship which can be recommended for the determination of the viscosity-average molecular weight of nylon 6 and 6,6 is $[\eta]$-M_w in m-cresol[5,23,109]. If it is necessary to make the estimate of M_n from the viscometric data, we regard it as best to start with a $[\eta]$-M_w relationship and to introduce a correction according to the probable or known values of the polydispersity index.

For the other nylons listed in Table 2 critical comparison cannot be performed because the data are scarce. A criterion of plausibility of a given $[\eta]$-M relationship may in this case consist in the comparison of the experimentally determined characteristic ratio of unperturbed dimensions, C_∞, with the theoretical value. The problem is discussed in detail in Section 5.3. Here it is only pointed out that the $[\eta]$-M_w relationships for nylon 4[4] and nylon 12[59–61] generally satisfy this criterion.

5.3 Conformational Properties

Basic information on the conformational rigidity of linear polymers is provided by the characteristic ratio of unperturbed dimensions, $C_\infty = \langle r_0^2 \rangle / nl^2$. Here $\langle r_0^2 \rangle$ is the mean square end-to-end distance of real chain, unperturbed by the volume effect, and nl^2 is the same quantity for a hypothetical freely jointed chain; the symbol n denotes the number of bonds in the main chain, l is the bond length. For simpler chains the characteristic ratio may be calculated theoretically. Flory[115] calculated C_∞, among others, for polyethylene (C_∞ = 8.0), nylon 6 (6.08), nylon 6,6 (6.10) and polyglycine, or nylon 2 (2.3). The decrease in C_∞ produced by introduction of relatively rigid amide groups into the polymethylene chain might seem pradoxical. The amide group, which has the character of a partial double bond with the predominant trans conformation, enhances in itself rigidity of the chain on the one hand, but at the same time releases rotation of methylene groups in its vicinity. The latter effect being stronger, the higher frequency of the amide group in the nylon chain raises its conformational flexibility.

The experimental value of the characteristic ratio is obtained from $[\eta]$ and M by means of the equation

$$C_\infty = (K_0/\Phi_0)^{2/3}\,(m/l^2), \tag{5}$$

where m is molecular weight related to one bond and Φ_0 is the Flory universal constant. The constant K_0 may be obtained directly from $[\eta]$ and M in the θ-conditions ($[\eta] = K_0 M^{0.5}$), or from data in thermodynamically good solvents by means of one of the extrapolation procedures[116], most often[117] extrapolation of the plot $[\eta]/M^{0.5}$ vs. $M^{0.5}$ to zero M. The C_∞ values in Table 2 have been cal-

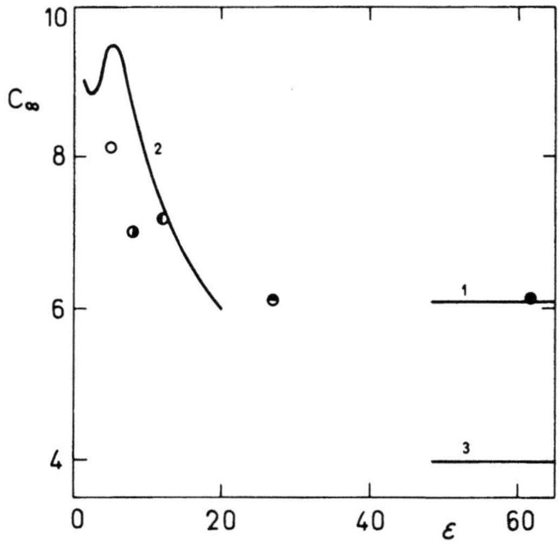

Fig. 14. Dependence of the characteristic ratio C_∞ on the dielectric constant ϵ of the solvent. ○ m-cresol/decalin[121], ◑ phenol/tetrachloroethane[121], ◐ m-cresol[5], ● TFP/0.1M LiCl/10 vol.-% H_2O[5], ● 85 wt.-% HCOOH[121]. 1 asymptote for $\epsilon \to \infty$ (nylon 6)[115], 2, 3 theoretical dependence and asymptote for $\epsilon \to \infty$ (polypeptides)[120], respectively[121]

culated only from the published K_o values; other values to be substituted in Eq. (5) are: $\Phi_o = 2.6 \times 10^{21}$, 1 = 2.17, 2.22, 2.24 and 2.28 Å for nylons 4, 6, 6,6, 8 and 12, respectively.

According to earlier views, unperturbed dimensions and thus also C_∞ should be independent of the nature of solvent. However, with a number of polymers (especially polar ones) an effect of the solvent has been observed[118, 119]. As was shown by Brand and Flory[120], interaction of dipoles of the amide groups in polypeptides affects rotation of the methylene groups to a great extent. They therefore predicted a decrease in the characteristic ratio with the dielectric constant of the solvent (Fig. 14). On the other hand, no such dependence was expected for polyamides, particularly those with a lower frequency of amide groups in the chain. However, experimental data[121] obtained with the same samples with a uniformly determined M_w also indicated the dependence of C_∞ on the dielectric constant of the solvent for nylon 6 (Fig. 14). Data of various authors may differ from each other because of experimental errors, the sources of which have been discussed above. In spite of this, the majority of the other available C_∞ values for nylon 6 and 6,6 do not seem to contradict the suggested dependence on the dielectric constant of the solvent.

The C_∞ values in m-cresol (having a lower dielectric constant) for nylons 4 and 12 are higher[4, 60, 61] than those in sulphuric acid[60] or alcoholic solvents[4, 122]; however, an attempt at a conclusive characterization of the dependence of C_∞ on the dielectric constant for these nylons would be hazardous, owing to the insufficient amount of experimental evidence.

An idea was also put forward[123] that the chain rigidity might be affected by specific bonds between polymer and solvent. Some authors[123] explained the high

C_∞ (9.14) of nylon 6,6 in methanol by complex formation between the amide group and calcium chloride in this solvent.

The temperature dependence of intrinsic viscosity of nylon 6 and 6,6 along with our own[121)] and literature[124, 125)] data were used in the determination of the temperature coefficient of the characteristic ratio $d\ln C_\infty/dT$. The values of this coefficient for the two nylons [121)] in m-cresol and 90 wt.-% HCOOH are very similar and vary between -3×10^{-3} and -4×10^{-3}. The fact that the amide bond has a partial double bond character[115)] implies the possibility of the cis-trans isomerism. It is probable that the equilibrium between the two forms is displaced in favor of the cis form with increasing temperature, which has as a consequence shortening of the nylon chain. The negative value of the temperature coefficient is in accordance with such a view. On the other hand, no temperature dependence of unperturbed dimensions was observed in a solution of nylon 6.6 in 85 wt.-% HCOOH by other authors[126)].

It follows from Flory's theoretical C_∞ values that the conformational rigidity of the nylon chain should decrease with increasing frequency of the amide groups in the chain. Similar to theoretical C_∞ values for nylon 6 and 6,6, experimental values do not differ on the average either, varying between 6 and 7. The C_∞ values of nylon 12 should be higher, *i. e.*, between those of nylon 6 (6.1) and polyethylene (8), while those of nylon 4 should be lower (ca. 5). Owing to scarce experimental data and the assumed error in C_∞ (ca. ± 1), such a tendency cannot yet be regarded as unambiguously proved.

6 Behavior in Mixed Solvents

The lack of one-component solvents of nylons and the specific requirements of the individual methods called for an application of a whole number of two- and multicomponent solvent systems. If the solvent components differ in their affinity to the polymer, one or several components of the solvent are selectively sorbed by the polymer[127)]. The composition of the solvent mixture in the region of the polymer coils is then different from that in the space between the coils. If the solvent components differ in their refractive index, the intensity of scattered light is determined by the effective refractive index increment, $(dn/dc)_\mu$. This parameter differs from the conventional refractive index increment, dn/dc, due to the effect of selective sorption. While dn/dc is determined from the difference of refractive indexes of the polymer solution and solvent, $(dn/dc)_\mu$ is given by difference between the refractive indexes of the polymer with the selectively sorbed solvent and the surrounding solvent. The latter difference is obtained, if the dialysis equilibrium is established between the polymer solution and solvent prior to the differential refractometric measurement. Thermodynamic reasoning[50)] shows that light scattering and sedimentation analysis yield true molecular weights of polymers dissolved in mixed solvents provided that $(dn/dc)_\mu$ is used instead of conventional dn/dc in the evaluation of experimental data. If dn/dc were substituted into the equations describing the light scattering or sedimentation phenomena, apparent molecular weights rather than true values would result.

The extent of selective sorption is quantitatively given by the coefficient of selective sorption, γ, defined as a change in the composition of the mixed solvent with a unit change in the polymer concentration at fixed chemical potentials of all solvent components[50, 129]. The coefficient of selective sorption can be obtained from the true and apparent molecular weight values or from the effective and conventional refractive index increments[128].

In systems polyelectrolyte/solvent/salt the Donnan equilibria usually cause displacement (desorption) ot the salt from the space occupied by polymer coils[129]. Surprisingly enough, in systems nylon 6 (or 6,6)/formic acid/salt none of the solvent components seems to undergo selective sorption, at least not to an extent which would make possible its determination from changes in the apparent molecular weight[13, 53, 54]. On the contrary, selective sorption of LiCl[4, 5] could be observed in the system nylon 6 (or 4)/2,2,3,3-tetrafluoropropanol/LiCl. Such an anomaly may be explained by a specific interaction between salt and polymer, which overcompensates the Donnan effect. According to some authors[130–132], lithium cations can form sufficiently strong complexes with the amide groups. This is also corroborated by the fact that in solutions of nylon 6 in 2,2,3,3-tetrafluoropropanol even a small concentration of LiCl in cooperation with water brings about dissolution of nylon aggregates[5], and that nylon 4 with a more amide groups sorbs more LiCl per weight unit than nylon 6 (Fig. 15)[4].

The dissolution of nylon 6 in methanolic $CaCl_2$[133] is also explained by formation of a strong bond between nylon and the methanol-$CaCl_2$ complex[123]. Infrared spectra suggest that Ca with the coordination number 6 forms a complex with alcohol molecules, able to solvate the carbonyl group of nylon[134] by one coordination valency (Fig. 16). The dissolving power of alcoholic solutions of other salts was also discussed from the viewpoint of complex formation[123, 135].

While addition of water to the alcoholic solution of $CaCl_2$ markedly lowers the dissolving power of the mixture for nylon 6, the presence of water in this solution

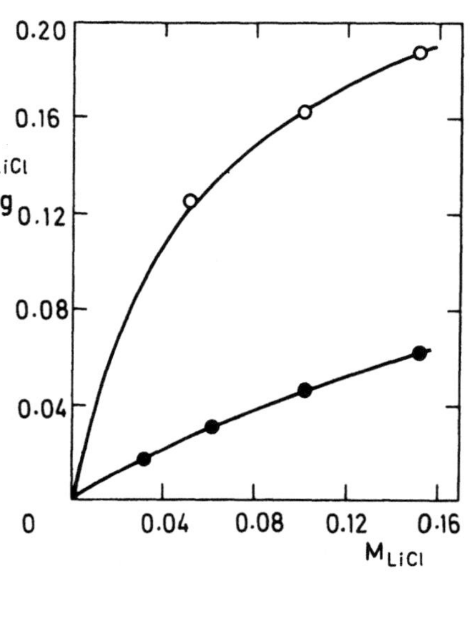

Fig. 15. Selective sorption coefficient of LiCl, γ_{LiCl}, on nylon 4 (○) and nylon 6 (●) from solutions of LiCl in TFP as a function of molarity of LiCl[4)]

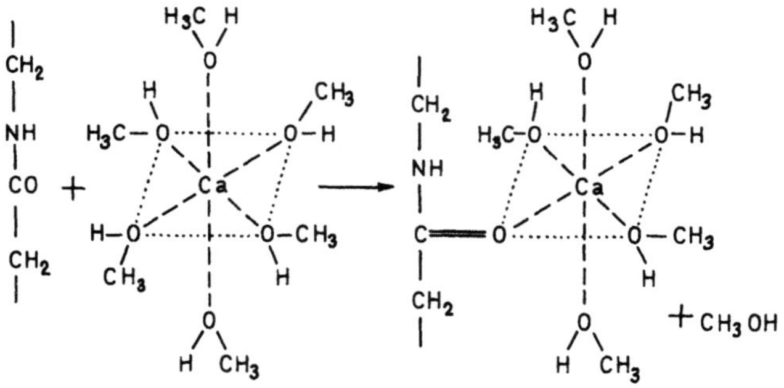

Fig. 16. Mechanism of dissolution of nylon 6 in methanolic $CaCl_2$[134)]

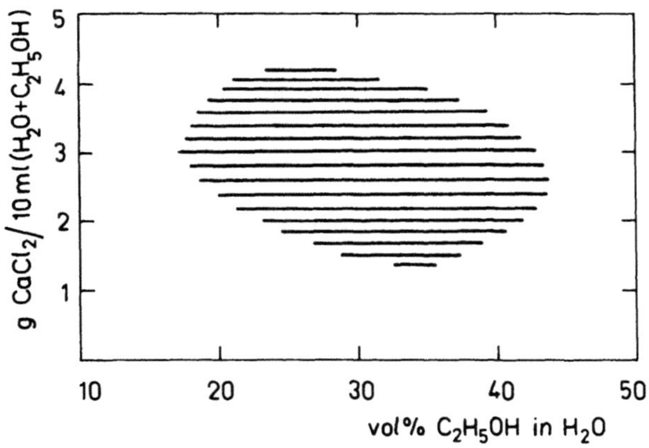

Fig. 17. Range of solubility of nylon 4 in the system $CaCl_2/H_2O/C_2H_5OH$[122)]

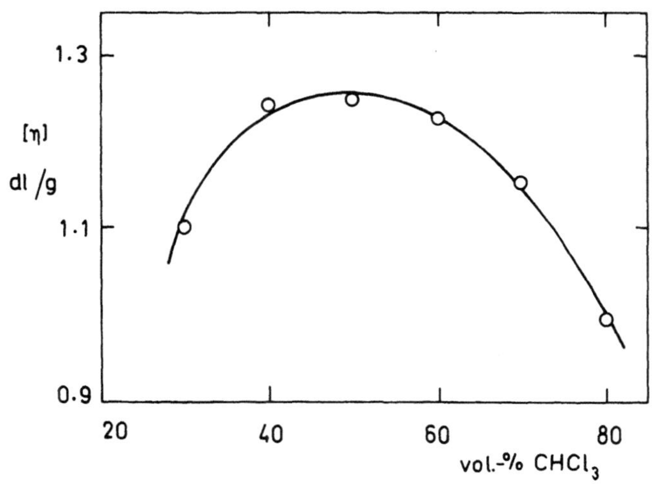

Fig. 18. Intrinsic viscosities of nylon 12 in the cosolvent mixture $TFP/CHCl_3$

is necessary if nylon 4 is to be dissolved[122] (Fig. 17). The active participation of $CaCl_2$ in the dissolution of nylon 4 was confirmed by the positive coefficient of selective sorption of $CaCl_2$[122].

A phenomenon called cosolvency, consisting in solution of a given polymer by two or more nonsolvents in a mixture has also been observed with nylon 12, namely, in mixtures 2,2,3,3-tetrafluoropropanol/30–80 vol.-% chloroform or 2,2,3,3-tetrafluoropropanol/45–60 vol.–% toluene[60]. A feature characteristic of cosolvency is the dependence of intrinsic viscosity on the composition of the cosolvent mixture (Fig. 18), where the maximum of the curve corresponds to the composition of the solvent with the relatively best thermodynamic quality. For such a solvent composition none of its components is selectively sorbed by the polymer[128].

With suitably chosen solvent components, the phenomenon of selective sorption can be used advantageously, especially in the light scattering method. The refractive index increment of a polymer in a single solvent is mainly determined by the difference between the refractive indexes of polymer and solvent. Table 1 shows that the highest value (0.26 ml/g) is exhibited by nylon 6 in 2,2,2-trifluoroethanol. If in a mixed solvent with components differing markedly in the refractive index values the component with the higher refractive index is strongly selectively sorbed on the polymer, the effective refractive index increment, $(dn/dc)_\mu$, may assume surprisingly high values which are no more limited by the difference between the refractive indexes of the polymer and solvent[62]. This has been observed, *e. g.* with nylons 4, 6 and 12 in the mixture m-cresol/n-heptane, where in the system nylon 6/m-cresol/50 vol-% n-heptane[62] a value $(dn/dc)_\mu = 0.816$ ml/g was obtained. It will probably be possible to make use of such systems in the determination of the M_w of nylons with a low degree of polymerization, *e. g.* oligomers.

7 Characterization of Branching

In the synthesis of nylons, branched molecules can in principle be formed by side reactions. Such a possibility is very real[136] for instance, with anionically polymerized nylons.

Statistical branching disturbs the regular structure of a linear chain and thus decreases the crystallinity of a branched polymer in the solid state. This unfavorably affects such properties as hardness or tensile strenght, but improves, *e. g.* impact or notch impact strength. Branching may also to some extent increase the heat resistance and shape stability. Statistical branching, controllable to a certain extent, may be achieved by, *e. g.* copolymerization of lactam with a small amount of bislactam[106].

Uncontrollable branching under conditions favouring side reactions, in the presence of a major amount of polyfunctional comonomers or induced by thermal treatment, may lead to the formation either of an insoluble fraction or of a small amount of strongly branched or cross-linked particles, often called microgels[137, 138]. These microgels, in spite of their low concentration, frequently have a decisive influence on mechanical properties (*e. g.* of fibres[137]). Model experiments yielded samples with a low content of microgels, the amount and size of which were determined by analyzing the angular dependence of the reduced intensity of scattered light[138].

While microgels are comparatively easy to detect in polymer solution, the determination of the degree of branching and often a mere proof of the existence of branching are experimentally difficult. In the investigation of statistical branching, the fact is utilized that a branched macromolecule with long branches occupies a smaller volume in the given solvent than a linear molecule of the same molecular weight, and that, with a change in the segment distribution in the domain of the macromolecule, the intramolecular hydrodynamic interaction also varies[139]. Hence, either the ratio of the radii of gyration, or more often intrinsic viscosities of the branched ($[\eta]_b$) and linear ($[\eta]_l$) polymer with the same molecular weight are used as the measure of branching[139]:

$$g' = [\eta]_b / [\eta]_l$$

where g' decreases with increasing degree of branching. A rigorous determination of the degree of branching requires measurements of $[\eta]_b$ and $[\eta]_l$ with identical molecular weight and molecular weight distribution in the θ-state; the condition of identical molecular weight distribution is virtually unrealizable. For a semiquantitative estimation of branching, data in a good solvent may also be employed[140].

A number of papers deal with the properties of intentionally branched nylons. Well-defined models are starlike branched nylons with four and eight branches synthetized from caprolactam in the presence of tetra- and octofunctional acids[113]. These polymers exhibit a rather narrow molecular weight distribution and a stable number of branches. Schaefgen and Flory[113] came to the surprising conclusion that a change in $[\eta]$ in concentrated H_2SO_4 appeared only in solutions of the polymer with eight branches; polymer with four branches behaved virtually identically with linear nylon 6. The starlike branching was more markedly reflected in the polyelectrolyte effect[104]: reduced specific viscosity of nylon 6 in 100 % HCOOH increased with decreasing concentration of both linear and branched samples. Since the dependences had no maxima[55], the authors[104] described them in terms of the Fuoss equation[141]

$$\eta_{sp}/c = A/(1 + Bc) + D \tag{5}$$

where A, B and D are constants for a given system; D was regarded as $[\eta]$ of an uncharged coil, and $(A + D)$ as $[\eta]$ of the polyion. The dependence of log $(A + D)$ on log M_n was linear for both linear and branched samples with four and eight branches[10] (Fig. 19) and showed that at the same M_n the intrinsic viscosity of the polyion demonstrably decreased with an increasing number of branches. The slope of the dependences (1.41 for a linear polymer, 1.45 for a polymer with four branches and 1.59 for a polymer with eight branches) suggested that chains (*i. e.* branches) are greatly stretched and that the effect of starlike branching was more pronounced at lower molecular weights.

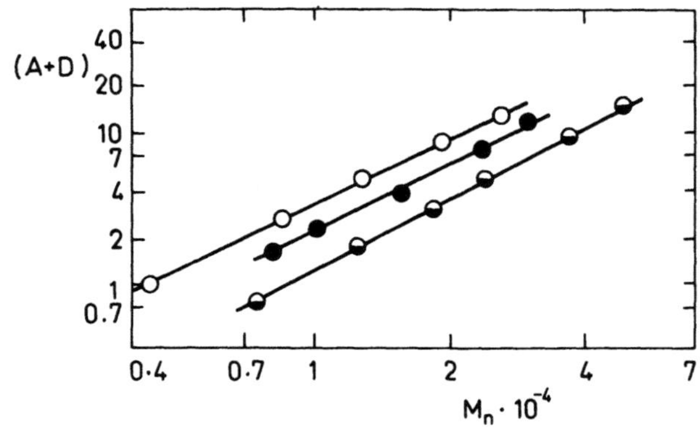

Fig. 19. Viscosity data of linear (○) and starlike nylon 6 samples with 4 (●) and 8 (⊖) branches in 100% HCOOH. For (A + D) see text[104)]

The method just described has clearly revealed branching of the hydrolytic nylon 6 prepared under conditions where formation of shorter branches might be expected[142)]. In the absence of the polyelectrolyte effect, the viscosity behavior of this sample did not differ from that of the linear polymer. The reliability of determination of branching based on the magnitude of the polyelectrolyte effect by comparing the intrinsic viscosities of the branched and linear polyion depends, of course, on the accuracy of determination of reduced viscosity in solutions where this effect is operative. It should also be borne in mind that results obtained with the system described above[104)] are very sensitive to the presence of even a tiny amount of water in formic acid, and that it is experimentally difficult to preserve the acid in the anhydrous state during viscometric measurements. The possibility of hydrolysis of nylon cannot be ruled out either.

A more detailed investigation of nylon 6 with eight branches in solutions without the polyelectrolyte effect in variously diluted formic acid showed that the dependence of the Mark-Houwink exponent a on the concentration of formic acid was different for branched and linear nylon; θ-conditions ($a = 0.5$) for the branched sample is reached at a lower concentration of formic acid[143)].

In the study of branching efficiency of various bislactams in the copolymerization with caprolactam, the extent of branching was characterized by the ratio of intrinsic viscosities of branched and linear nylon 6 at the same M_w[106)], *i. e.* by g'. Comparison of the dependence $[\eta]$-M_w measured with fractions of the branched sample with the dependence for linear nylon 6 (Fig. 20) shows that the branching ratio g' decreases with increasing M_w of the fraction. This result is in agreement with the assumption of the statistical mechanism of branching[106)]; similar dependences have also been frequently described for the copolymerization products of vinyl monomers with a small amount of tetrafunctional comonomers[138)].

The branching ratio g' may be affected to some extent by the polydispersity of the sample: The higher the latter, the lower $[\eta]$, even if $[\eta]$ is related to M_w. This effect is rather weak, but might become strongly operative, if $[\eta]$ were related to the number-average molecular weights. If the weight-average molecular weights and intrinsic viscosities in a thermodynamically good solvent are used, the poly-

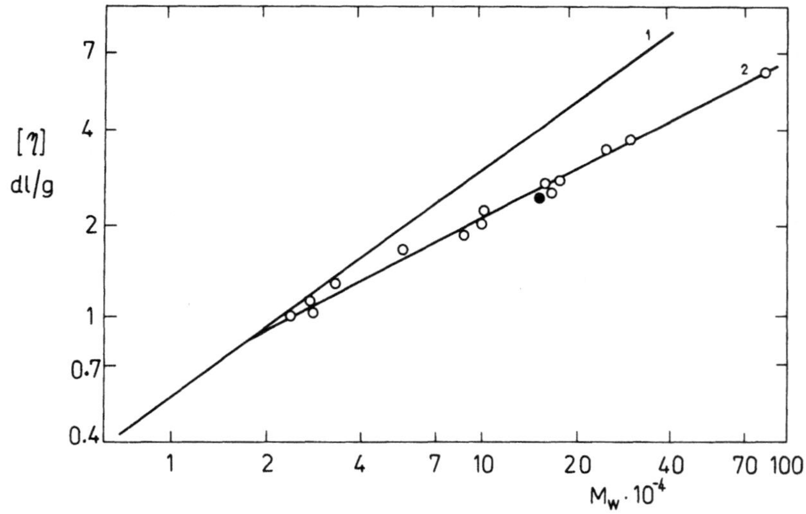

Fig. 20. [η]-M_w dependences for linear (1) and branched (2) nylon 6. ○ fractions, ● unfractionated sample[106]

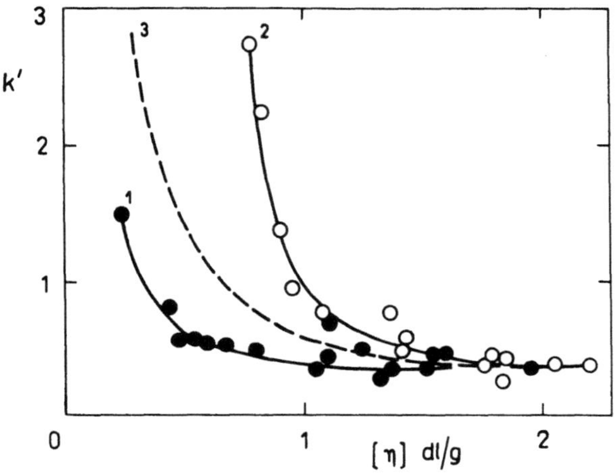

Fig. 21. Dependence of the Huggins coefficient, *k'*, on intrinsic viscosity of nylon 6. 1 branched polymer with only carboxyl end groups, 2 branched polymer with both carboxyl and amino end groups, 3 linear polymer[145]

dispersity effect seems to be negligible: In the relationship [η]-M_w for both linear and branched nylon 6 (Fig. 20) the effect of the polydispersity of unfractionated samples is not operative, *i. e.* unfractionated samples and their fractions fit the same dependence. To achieve the highest possible accuracy in the determination of *g'*, it is advisable to measure the M_w of branched samples by the same method and under the same conditions as the M_w of linear polymers.

Goebel[107] determined the average number of branches in the molecule of branched nylon 6 from the osmometrically measured M_n and from the overall concentration of end groups determined by chemical analysis. This is a direct method, which under favorable conditions, permitting a sufficiently accurate determination both of end groups and of the osmometric M_n, may provide plausible results. A probably less exact estimate of the average number of branches is based on the comparison of the content of end groups before and after hydrolysis of branched nylon with sulphuric acid. The method is based on the assumption of a lower stability, and thus of the selective cleavage of the imide and imidine branching sites; it was applied to the interfacially polymerized nylon 6,10[144].

In the literature, the magnitude of the Huggins coefficient *k'* (*e. g.* for nylon 6 in dilute formic and sulphuric acid[56]) is also given as a criterion of branching. However, any unequivocal conclusions about the branching of nylons based on the higher *k'*values are risky because of the sensitivity of this coefficient to a number of other factors, particularly to the character of the end groups[145]. While *k'* varies only insignificantly with $[\eta]$ in the case of the starlike branched nylon 6, which contain exclusively carboxyl end groups (Fig. 21, curve 1), it increases sharply with decreasing $[\eta]$ for polymers branched with various bislactams containing both carboxyl and amine end groups (curve 2). The difference between curves 1 and 2 suggests that the effect of association of carboxyl and amine groups on *k'*may be larger than the effect of branching.

Of the above methods of estimation of branching, the determination of *g'*as the ratio of intrinsic viscosities of the branched and linear sample in m-cresol at the same M_w seems to be fairly reliable. It should be borne in mind, that a low degree of branching need not be reflected in this procedure either. In this case a qualitative proof of branching may be based on the fact that branching increases with molecular weight[106], and *g'*may be determined for an isolated high-molecular weight fraction of such sample. A convincing, though time-consuming, proof of branching consists in the comparison of the $[\eta]$-M_w relation of fractions of nylon with that for the corresponding linear nylon[106].

8 References

1. Burke, J. J., Orofino, T. A.: J. Polym. Sci. *A-2, 7,* 1 (1969)
2. Valentine, L.: J. Polym. Sci. *23,* 297 (1957)
3. Vrhel, Z.: Thesis, Technical University, Prague 1974, (in Czech)
4. Tuzar, Z., Kletečková, J., Králíček, J.: Colln Czech. chem. Commun. *39,* 2206 (1974)
5. Tuzar, Z., Kratochvíl, P., Bohdanecký, M.: J. Polym. Sci. *C 16,* 633 (1967)
6. Tuzar, Z.: unpublished
7. Heikens, D.: J. Polym. Sci. *35,* 277 (1959)
8. Matthes, A.: J. prakt. Chem. *162,* 245 (1943)
9. Matthes, A.: Makromol. Chem. *5,* 165 (1950)
10. Schaefgen, J. R., Trivisonno, C. F.: J. Am. Chem. Soc. *73,* 4580 (1951)
11. Liquori, A. M., Mele, M.: Gazz. chim. it. *82,* 828 (1952)
12. Hoshino, K., Watanabe, M.: J. Am. Chem. Soc. *73,* 4816 (1951)
13. Tuzar, Z., Kratochvíl, P.: Colln Czech. chem. Commun. *32,* 2255 (1967)
14. Benewitz, R.: Faserforsch. Textiltechn. *5,* 155 (1954)
15. Bonnar, R. U., Dimbat, M., Stross, F. H.: Number average molecular weights. New York: Intersci. Publ. Inc. 1958

16. Garmon, R. G.: End group determinations. In: Techniques and methods of polymer evaluation, Vol. 4. Polymer molecular weights, Part I. Slade, P. E., Jr. (ed.). New York: M. Dekker, Inc. 1975
17. Huglin, M. B. (ed.): Light scattering from polymer solutions. London: Academic Press 1972
18. Carothers, W. H., Van Natta, F. J.: J. Am. Chem. Soc. *55*, 4714 (1933)
19. Garmon, R. G., Gibson, M. E.: Anal. Chem. *37*, 1309 (1965)
20. Dawydoff, W. N.: Bestimmung des Molekulargewichts von Polyamiden. Berlin: VEB Verlag Technik 1954
21. Broser, W.: Makromol. Chem. *2*, 248 (1948)
22. Basu, S.: J. Polym. Sci. *5*, 735 (1950)
23. Reimschuessel, H. K., Dege, G. J.: J. Polym. Sci. *A-1*, *9*, 343 (1971)
24. Waltz, I. E., Taylor, G. B.: Anal. Chem. *19*, 448 (1947)
25. Griehl, W., Sieber, H.: Faserforsch. Textiltechn. *6*, 329 (1955)
26. Wang, Y., Chen, K., Chang, T., Yang, H., Chien, J., Wang, P.: Acta Chim. Sinica, *22*, 305 (1956)
27. Elias, H.-G., Schumacher, R.: Makromol. Chem. *76*, 23 (1964)
28. Wolf, S., Mobus, B.: Z. Anal. Chem. *186*, 194 (1962)
29. Pätzold, W.: J. Polym. Sci. *B*,*1*,269 (1963)
30. Loepelman, F.: Faserforsch. Textiltechn. *4*, 58 (1952)
31. Maurice, M. J.: Anal. Chim. Acta *26*, 406 (1962)
32. Dinse, H., Praeger, K.: Faserforsch. Textiltechn. *21*, 305 (1970)
33. Stea, G., Gechele, G. B.: Eur. Polym. J. *1*, 213 (1965)
34. Head, F. S. H.: J. Polym. Sci. *A-1*, *7*, 2456 (1969)
35. Head, F. S. H.: J. Polym. Sci. *A-1*, *8*, 3651 (1970)
36. Reimschuessel, H. K., Dege, G. J.: J. Polym. Sci. *A-1*, *8*, 3265 (1970)
37. Lánská, B., Šebenda, J.: Colln Czech. chem. Commun. *39*, 257 (1974)
38. Hoshino, K.: Bull. Chem. Soc. Japan *19*, 158 (1944)
39. Batzer, H., Möschle, A.: Makromol. Chem. *22*, 195 (1957)
40. Bruck, S. D., Bair, H. E.: Polymer *6*, 447 (1965)
41. Biernacki, P., Wlodarczyk, M.: Eur. Polym. J. *11*, 107 (1975)
42. Ohama, M., Ozawa, T.: J. Polym. Sci. *A-2*, *4*, 817 (1966)
43. Nichols, J. B.: Colloid aspects of nylon. In: Colloid chemistry, Vol. 6 (1946), Alexander, J., Jr. (ed). New York: Reinhold 1949
44. Schumacher, R., Elias, H.-G.: Makromol. Chem. *76*, 12 (1964)
45. Taylor, G. B.: J. Am. Chem. Soc. *69*, 635 (1947)
46. Provder, T., Woodbrey, J. C., Clark, J. H.: Separ. Sci. *6*, 101 (1971)
47. Goebel, C. V.: Forth Internat. GPC Seminar, Miami Beach, Florida, Seminar Proc. *1967*, 90
48. Huglin, M. B., Richards, R. W.: Angew. Makromol. Chem. *38*, 207 (1974)
49. Beachell, H. C., Carlson, D. W.: J. Polym. Sci. *40*, 543 (1959)
50. Casassa, E. F., Eisenberg, H.: Adv. Protein Chem. *19*, 287 (1964)
51. Fendler, H. C., Stuart, H. A.: Makromol. Sci. *25*, 159 (1957)
52. Saunders, P. R.: J. Polym. Sci. *43*, 273 (1960)
53. Saunders, P. R.: J. Polym. Sci. *57*, 131 (1962)
54. Saunders, P. R.: J. Polym. Sci. *A*,*2*, 3755 (1964)
55. Quadrat, O., Bohdanecký, M.: Colln Czech. chem. Commun. *29*, 2449 (1964)
56. Sbrolli, W., Capaccioli, T.: Chim. Ind. (Milan) *42*, 3 (1960)
57. Saunders, P. R.: J. Polym. Sci. *A*,*3*, 1221 (1965)
58. Matzner, M., Robeson, L. M., Greff, R. J., McGrath, H. E.: Angew. Makromol. Chem. *26*, 137 (1972)
59. Hammel, R., Gerth, Ch.: Makromol. Chem. *178*, 2697 (1977)
60. Tuzar, Z., Bohdanecký, M., Puffr, R., Šebenda, J.: Eur. Polym. J. *11*, 851 (1975)
61 Lánská, B., Bohdanecký, M., Šebenda, J., Tuzar, Z.: Eur. Polym. J., *14*, 807 (1978)
62. Tuzar, Z., Kratochvíl, P.: Macromolecules *10*, 1108 (1977)
63. Taylor, G. B.: J. Am. Chem. Soc. *69*, 638 (1947)

64. Wiloth, F.: Makromol. Chem. *14*, 156 (1954)
65. Laczkovski, M., Turska, E.: Roczniki chimii *29*, 941 (1955)
66. Turska, E., Laczkowski, M.: J. Polym. Sci. *23*, 285 (1957)
67. Korshak, V. V., Zamyatina, V. A.: Dokl. Akad. Nauk. S.S.S.R. *59*, 909 (1948)
68. Gordienko, A.: Faserforsch. Textiltechn..*4*, 499 (1953)
69. Howard, G. J.: J. Polym. Sci. *37*, 310 (1959)
70. Biernacki, P., Chrzczonowicz, S., Wlodarczyk, M.: Eur. Polym. J. *7*, 739 (1971)
71. Duveau, N., Piguet, A.: J. Polym. Sci. *57*, 357 (1962)
72. Gechele, G. B., Mattiussi, A.: Eur. Polym. J. *1*, 47 (1965)
73. Mattiussi, A., Manescalchi, F., Gechele, G. B.: Eur. Polym. J. *5*, 105 (1969)
74. Rafler, G., Reinisch, G.: Faserforsch. Textiltechn. *21*, 91 (1970)
75. Rafler, G., Reinisch, G.: Angew. Makromol. Chem. *20*, 57 (1971)
76. Griehl, W., Lückert, H.: J. Polym. Sci. *30*, 399 (1958)
77. Threlkeld, J. O., Ende, H. A.: J. Polym. Sci. *4*, 663 (1966)
78. Flory, P. J.: J. Am. Chem. Soc. *58*, 1877 (1936)
79. Hoshino, K., Watanabe, M.: J. Chem. Soc. Japan, Pure Chem. Sect. *70*, 24 (1949)
80. Bohdanecký, M., J. Polym. Sci. *C,23*, 257 (1968)
81. Juiffs, J.: Kolloid-Z. *141*, 88 (1955)
82. Froyer, G., Sekiguchi, H.: C. R. Acad. Sci. Paris t 284, C-525 (1977)
83. Craubner, H.: J. Polym. Sci. *C, 42*, 889 (1973)
84. Urwin, J. R.: Molecular weight distributions by turbidimetric titration, In: Ref.[17], Chapter 18, p.789
85. Howard, G. J.: J. Polym. Sci. *A,1*, 2667 (1963)
86. Gordienko, A., Griehl, W., Sieber, H.: Faserforsch. Textiltechn. *6*, 105 (1955)
87. Gehrke, K.: Faserforsch. Textiltechn. *17*, 524 (1966)
88. Conti, W., Sorta,: Eur. Polym. J. *8*, 475 (1972)
89. Dudley, M. A.; J. Appl. Polym. Sci. *16*, 493 (1972)
90. Provder, T., Rosen, E. M.: Separ. Sci. *5*, 437 (1970)
91. Ede, P. S.: J. Chromatogr. Sci. *9*, 275 (1971)
92. Panaris, R., Pallas, G.: J. Polym. Sci. *B, 8*, 441 (1970)
93. Huggins, M. L.: J. Am. Chem. Soc. *64*, 2716 (1942)
94. Kraemer, E. O.: Ind. Eng. Chem. *30*, 1200 (1938)
95. Bohdanecký, M., Lánská, B., Šebenda, J., Tuzar, Z.: Eur. Polym.J. in press
96. Heller, W.: J. Colloid Sci. *9*, 547 (1954)
97. Martin, A. F.: Am. Chem. Soc. Meeting, Memphis 1942
98. Staudinger, H., Heuer, W.: Z. Phys. Chem. *A 171*, 129, (1934)
99. Bohdanecký, M.: unpublished
100. Rybnikář, F.: Colln Czech. chem. Commun. *21*, 1101 (1956)
101. Lánská, B., Šebenda, J.: Chem. Průmysl (in Czech), in press
102. Šebenda, J., Králíček, J.: Colln Czech. chem. Commun. *31*, 2534 (1966)
103. Rybnikář, F.: Kolloid-Z. *158*, 148 (1958)
104. Schaefgen, J. R., Trivisonno, C. F.: J. Am. Chem. Soc. *74*, 2715 (1952)
105. Kurata, M., Tsunashima, Y., Iwama, M., Kamada, K.: Viscosity molecular weight relationships and unperturbed dimensions of linear chain molecules, in Ref.[111]
106. Kondelíková, J., Tuzar, Z., Králíček, J., Šandová, K., Strohalmová, M., Kubánek, V.: Angew. Makromol. Chem *64*, 123 (1977)
107. Goebel, C. V., Čefelín, P., Stehlíček, J., Šebenda, J.: J. Polym. Sci., Polym. Chem. Ed. *10*, 1411 (1972)
108. Hague, P., Huglin, M. B., Johnson, B. L., Smith, J.: J. Appl. Polym. Sci.*12*, 2105 (1968)
109. Kulakova, A. M., Nekrasov, I. K.: Vysokomol. soed. *A, 18*, 127 (1976)
110. Pavlov, A. V., Bresler, S. Ya., Rafikov, S. R.: Vysokomol. soed. *6*, 2068 (1964)
111. Bareiss, R. E.: Polymolecularity correction factors. In: Polymer handbook, Brandrup, J., Immergut, E. H. (eds.). New York: John Willy and Sons, Inc. 1975
112. Fukumoto, O.: J. Polym. Sci. *22*, 263 (1956)
113. Schaefgen, J. R., Flory, P. J.: J. Am. Chem. Soc. *70*, 2709 (1948)
114. Saunders, P. R.: J. Polym. Sci. *A, 2*, 3765 (1964)

115. Flory, P. J., Williams, A. D.: J. Polym. Sci. *A-2, 5,* 399 (1967)
116. Berry, G. C.: J. chem. Phys. *46,* 1338 (1967)
117. Stockmayer, W. H., Fixman, M.: J. Polym. Sci. *C 1,* 137 (1963)
118. Flory, P. J., Crescenzi, V., Mark, J. E.: J. Am. Chem. Soc. *86,* 146 (1964)
119. Ivin, K. J., Ende, H. A., Meyerhof, G.: Polymer *3,* 129 (1962)
120. Brandt, D. A., Flory, P. J.: J. Am. Chem. Soc. *87,* 2791 (1965)
121. Bohdanecký, M., Tuzar, Z.: Colln Czech. chem. Commun. *34,* 2589 (1969)
122. Tuzar, Z., Kvarda, V., Králíček, J.: Colln Czech. chem. Commun. *42,* 1514 (1977)
123. Nakajima, A., Tanaami, K.: Polymer J. *5,* 248 (1973)
124. Liquori, A. M., Mele, M.: J. Polym. Sci. *13,* 589 (1954)
125. Rybnikář, F.: J. Polym. Sci. *29,* 519 (1958)
126. Mattiussi, A., Gechele, G. B., Francesconi, R.: J. Polym. Sci. *A-2, 7,* 411 (1969)
127. Ewart, R. H., Roe, C. P., Debye, P., McCartney, J. R.: J. Chem. Phys. *14,* 687 (1964)
128. Tuzar, Z., Kratochvíl, P.: Colln Czech. chem. Commun. *32,* 3358 (1967)
129. Vrij, A., Overbeek, J. T. G.: J. Colloid Sci. *17,* 570 (1962)
130. Mandelkern, L., Halpin, J. C., Diorio, A. F., Posner, A. S.: J. Am. Chem. Soc. *84,* 1383 (1962)
131. Sarda, G., Peacock, N.: Nature *200,* 67 (1963)
132. Hamoud, M. R., Vert, M., Selegny, E.: J. Polym. Sci. *B, 10,* 361 (1972)
133. Hargreaves, G.: J. Polym. Sci. *B,2,* 29 (1964)
134. Aoki, T.: Lecture Meeting "Technique for Producing Synthetic Leather", Japanese Soc. Mechanical Engineering, Tokyo, March 19, 1959, quoted in[123)]
135. Ford, R. A., Marshall, H. S. B.: J. Polym. Sci. *22,* 351 (1956)
136. Bukač, Z., Šebenda, J.: Colln Czech. chem. Commun. *32,* 3537 (1967)
137. Jaeger, W., Jacob., Reinisch, G.: Faserforsch. Textiltechn. *21,* 529 (1970)
138. Dietrich, W., Basch, A.: Angew. Makromol. Chem. *40/41,* 159 (1974)
139. Small, P. A.: Adv. Polym. Sci. *18,* 1 (1975)
140. Krigbaum, W. R., Trementozzi, Q. A.: J. Polym. Sci. *28,* 295 (1958)
141. Fuoss, R. M., Cathers, G. I.: J. Polym. Sci. *4,* 97 (1949)
142. Mattiussi, A., Gechele, G. B., Pizzoli, N.: Eur. Polym. J. *2,* 383 (1966)
143. Francesconi, R., Gechele, G. B., Mattiussi, A.: Eur. Polym. J. *5,* 273 (1969)
144. Morgan, P. W., Kwolek, S. L.: J. Polym. Sci. *A,1,* 1147 (1963)
145. Králíček, J., Kondelíková, J., Kubánek, V.: Scientific papers of the Institute of chemical technology, Prague *1972,* C 18
146. Mermond, Ch.: Bull. Soc. Chim. France *1961,* 1614
147. Sekiguchi, H.: Bull, Soc. Chim. France *1960,* 1839
148. Coutin, B.: Thesis, Université Paris, 1974
149. Rusznak, I., Geczy, I., Ady, E.: Faserforsch. Textiltechn. *7,* 490 (1956)
150. Chien Jen-Yuan, Shih Liang-Ho, Shih Kwan-I: Acta Chim. Sinica, *21, 50* (1955)
151. Reinish, G.: Faserforsch. Textiltechn. *25,* 111 (1974)
152. Ogata, O.: Bull. Soc. Chem. Japan, *33,* 1584 (1960)
153. Burnett, G. M., MacArthur, A. J. Hay, J. N.: Eur. Polym. J. *3,* 321 (1967)
154. Okada, R., Fukumura, T., Tanzawa, H.: J. Polym. Sci. *B,4,* 971 (1966)
155. Howard, G. J.: J. Polym. Sci. *39,* 548 (1959)
156. Manescalchi, F., Mattiussi, A.: Atti Mem. Accad. Potravina Sci., Lett. Arti, Classe Sci. Mat. Nat. *84,* 275 (1972)
157. Puffr, R., Šebenda, J.: J. Polymer Sci. *C, 42,* 21 (1973)
158. Griehl, W., Zarate, J.: Plastverarbeiter *18,* 527 (1967)

Received June 5, 1978
K. Dušek (editor)

A General Theory for the Evaluation of X-Ray Diagrams of Biomembranes and Other Lamellar Systems

Wolfram Welte and Werner Kreutz

Institut für Biophysik und Strahlenbiologie, der Universität Freiburg i. Br., Albertstr. 23, D-7800 Freiburg, Germany

In the following contribution the development of a new evaluation theory for X-ray diffraction diagrams of biomembranes and other lamellar systems is presented. This theory is more comprehensive than other theories applied hitherto, since statistical lattice distortions, distortions of the lattice-cell, membrane undulations and different interspatial electron densities as well as "Lorentz" correction are considered.

Table of Contents

1 Fundamentals of Small-Angle X-Ray Scattering Theory of Membrane Stacks

1.1 Introduction

The problem of small-angle X-ray scattering of lamellar systems will be treated in the following chapters mainly by means of biomembrane aspects. Nevertheless, the concepts should generally apply to other lamellar systems, such as polymer systems.

Special aspects apply to biomembranes in so far as these lamellar objects always occur in a topologically closed form, called *vesicles.* In a *biomembrane stack* a number of vesicles are flattened and stacked along their short axes. See Fig. 1 for an example of such biomembrane stacks and Fig. 2 for an example for a scattering diagram of an ensemble of such stacks. Thinking in crystallographic terms about biomembrane stacks, we conclude from the topological closure of the single vesicles that the electron density profile of the cross section of the unit cell always has a centre of symmetry. We will see, however, that the centre-to-centre distance of the two unit cell membranes as well as the centre-to-centre distance of adjacent unit cells show statistical variations. The only invariable structure in biomembrane stacks is therefore the single membrane (Fig. 1).

The purpose in writing this article was on the one hand to give an historical survey of current literature on theoretical methods and their applications in small-angle diffraction of biomembrane stacks and on the other hand to present a clear and logical description of the basic principles of evaluating the X-ray scattering dia-

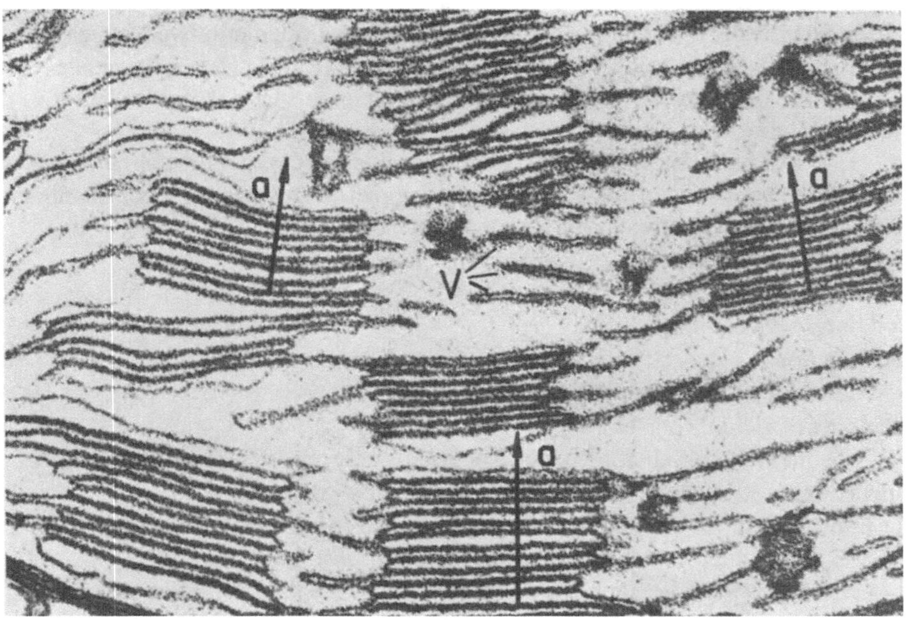

Fig. 1. Electron microscopic picture of membrane stacks in photosynthetic centres (chloroplasts) of higher plants with slightly varying orientation of their stacking axes (*a*) (Flower Bunch Statistics). The membrane stacks consist of stapled membrane vesicles (*v*) (closed double membranes)

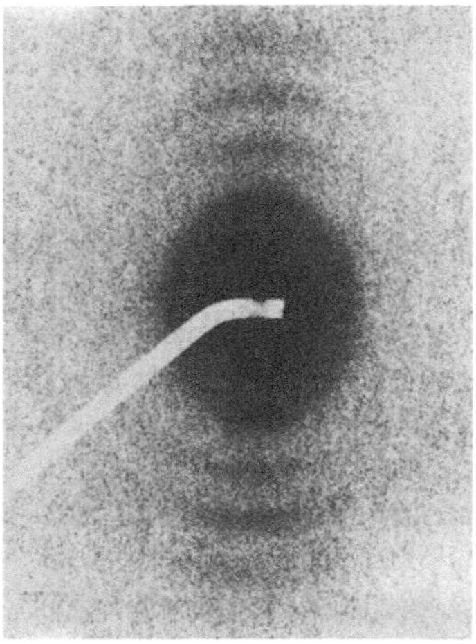

Fig. 2. Meridional X-ray diffraction pattern of preferentially oriented stacks of photosynthetic membranes of bacteria (Rh. viridis), showing distance statistics, membrane undulations and different interspace electron densities[47]

grams of such systems. The two points of view seemed to be incompatible. We therefore decided to emphasize a clear derivation of a comprehensive theory and its assumptions, which will not follow the historical development. This presentation will nevertheless build upon the various theoretical attempts found in the literature and ultimately will take into account two aspects which have been neglected up to today: the possibility of different electron density levels in the two types of intermembranous spaces (Chap. 3) and the partial correlations of undulated membrane surfaces in the biomembrane stack (Chap. 4).

Before beginning with the systematic theory in Chap. 1, we will give a short historical view of the literature. Literature can be subdivided into two main groups. The first of these can be termed *crystalline*, for its roots lie in classical crystallography. The other can be termed *paracrystalline*. Its roots lie in X-ray studies of liquids[1,2] and gases.

The crystalline studies deal with long, crystalline stacks of biomembrane vesicles, the unit cell consisting of the profile of two symmetrically arranged single membranes. Biomembrane stacks investigated in this way are stacks of:

lipid bilayers[3–11]
nerve myelin[12–14]
rod outer segment discs[15, 16]
gap-junctions[17]
several membranes of Halobacterium halobium[18–20]
photosynthetic membranes[21]

The method of evaluation of the electron density, which was most frequently applied in these studies, consists of the following two steps:

1) Evaluation of the structure factor (equivalent to the autocorrelation function or Q_0-function of a single unit cell by Fourier-transformation, 2.1).

The integrated reflexions of a crystalline arrangement of unit cells with period L represent a set of equidistant points (with reciprocal distance 1/L) of the structure factor[22]. Changing L by swelling without affecting the unit cell profile, provides a different set of equidistant points of the structure factor, so that eventually the whole structure factor can be measured.

2) Evaluation of the density profile of the unit cell from the structure factor.

Here all the methods apply which are described in 2.2. See also[23]. A good survey of this method is given in[13, 24].

All crystalline studies deal with narrow reflexions (due to the periodic arrangement of many unit cells) or correspondingly with infinite Patterson functions. In contrast to this, the scope of the present article consists of scattering intensities with broad reflexions or correspondingly short Q-functions (generalized Patterson functions) with distance correlations only within some 1–4 Bragg-periods, since this case gives more than just one set of equidistant points of the structure factor by a single measurement and consequently more information about the unknown phases within one measurement. This can occur either in the case of small crystalline stacks of biomembranes, consisting of only 1–4 unit cells, or in the case of paracrystalline stacks. The previous case of short, crystalline stacks of biomembranes (Chap. 2) has had only a limited number of applications[25–28]. Nevertheless, we reach here some important conclusions with regard to the information contained in broad Bragg-reflexions or short Q-functions (s. Chap. 2). Some literature of this group uses scattering from only single unit cells in order to measure the structure factor directly[5, 6, 29].

The paracrystalline part of literature begun with investigations of lattice distortions in liquid systems[1, 2]. Two important mathematical concepts emerged, *distortions of the first* and *of the second kind.* Another concept was introduced by studies of cellulose micelles; in these objects the unit cell itself showed statistical variations. It was therefore necessary to allow for statistical variations of the unit cell[30].

Ewald[31] was the first to use convolution operations in the scattering theory. The fruitful concept of autocorrelation functions thus appeared. Calculations of the autocorrelation function, called Q-function, proved to be more adequate to paracrystalline problems than intensities. Hosemann's perfect completion of the theory[32] used the Q-function extensively.

The first direct (non trial and error) method to phase a structure amplitude of a symmetrical unit cell was found with the help of its autocorrelation or Q_0-function[32c]. Pape's direct deconvolution procedure[33] (see 2.3) is also based on the Q_0-function.

One of the authors of this article has introduced the paracrystalline concept of above[34–36] and also the Q-function method[36] into the evaluation of scattering diagrams of biomembrane stacks. The paracrystalline and Q-function approaches were continuously pursued and completed in the following years[37, 38, 40, 41, 43, 48, 55].

Mainly two models for biomembrane stacks were defined in the following by paracrystalline concepts, characterized by specific *assumptions of the model:* the *vesicle concept* and the *membrane concept.*

With regard to the lattice distortions, all authors agree that second kind distortions should be the appropriate model for biomembrane stacks.

A one-dimensional paracrystal with lattice distortions of the second kind is a series of unit cells which may not be identical and whose distances follow a nearest neighbour distribution function. Some authors[34–37, 39] transcribed this definition to biomembrane stacks by defining each biomembrane vesicle as a unit cell. In this *vesicle concept* a lattice with distortions of the second kind fixed the position of the membrane vesicles and the statistical character of the unit cell was transferred to the statistical variation of the intravesicular intermembrane distance[34–37,39]. In the other approach the concept of a unit cell was abandoned[38–41, 43, 48]. It was replaced by two alternating, uncorrelated nearest neighbour distribution functions to build up the lattice, each of these defining the localisation of a certain membrane of the two possible ones (*membrane concept*, s. 3.1).

Nearly all authors use gaussians as analytical model functions of the nearest neighbour distributions, though all agree that this is only an approximation to the real distribution[41].

In all paracrystalline studies, with the exception of the present one, the electron densities of the intermembranous spaces are assumed to be equal to the mean electron density of the biomembrane stacks.

Allowing for different intermembranous electron density levels results in new contributions to the scattering intensity. One contribution is a background like scattering, which can be very prominent, when the mean electron density of the membrane differs appreciably from that of the two intermembranous spaces.

Some authors discuss the effect of perfectly correlated membrane undulations on the scattering diagram[42]. We improve this approach by adopting the general case of partially correlated membrane undulations (4.3).

When discussing the evaluation of electron density profiles of other paracrystalline studies, one has to keep in mind besides the assumptions of the models the special methods they use to determine the electron density profile from the experimental data. In the paracrystalline literature these methods are either very intricate[37] or poorly described[41]. We therefore concentrate on our own methods (3.2, 4.5), which seem to us to be the conceptually simplest of all existing methods.

The following types of biomembrane stacks were investigated by paracrystalline methods:
nerve myelin[41,43]
retinal rod outer segment discs[37]
human erythrocyte ghost membranes[38]
sarcoplasmic reticulum[44]
chromatophore membrane[45]
cytochrome oxidase model membranes[46]
photosynthetic membranes[34–36, 47]
mitochondrial cristae membranes[48]

The following treatise presupposes a general knowledge of the X-ray diffraction theory, especially of small-angle X-ray diffraction and familarity with the laws of Fourier-optics. Pertinent information may be obtained from the book of Hosemann and Bagchi[32]. An introductory presentation of this theory is also given in the summer school lectures on the Biophysics of Membranes held in Vancouver in 1974 by Kreutz and Pape[39].

1.2 X-Ray Diffraction Theory for Small Angles

X-ray scattering can be described by the laws of Fourier-optics[32a] which are summarized in the following diagram:

$$\begin{array}{ccc} I(\vec{b}) & \xleftarrow{\;||^2\;} & F(\vec{b}) \\ F\uparrow & & F\uparrow \\ Q(\vec{x}) & \xleftarrow{\;\underset{\sim}{2}\;} & \rho(\vec{x}) \end{array} \qquad (1.1)$$

$\rho(\vec{x})$ is the electron density, $F(\vec{b})$ the scattering amplitude, F abbreviates the Fourier transform,

i.e. $F(\vec{b}) = F\rho(\vec{x}) \equiv \int \rho(\vec{x})\, e^{-2\pi i \vec{b}\vec{x}}\, d\tau$

$Q(\vec{x})$ is the autocorrelation function or convolution square of $\rho(\vec{x})$. Physically, $Q(\vec{x})$ is the frequency distribution of the distances connecting all possible pairs of electrons in the sample, i.e. it has a large value, if there is a large number of electrons in the sample, that can be connected by the vector $\vec{x}$. The symbol $\underset{\sim}{2}$ means *convolution square,* a special kind of the
Convolution of two arbitrary functions $a(\vec{x})$ and $g(\vec{x})$ is defined by:

$$h(\vec{x}) \equiv a(\vec{x}) * g(\vec{x}) \equiv \int a(\vec{y})\, g(\vec{x} - \vec{y})\, d\tau_y \qquad (1.2)$$

The symbol $*$ abbreviates the convolution integral. The Q-function, the *convolution square* of $\rho(\vec{x})$ is identical with the convolution product of $\rho(\vec{x})$ with its mirror image:

$$\rho^-(\vec{x}) \equiv \rho(-\vec{x})$$

$$Q(\vec{x}) = \int \rho(\vec{y})\, \rho(\vec{y} - \vec{x})\, d\tau_y \equiv \overset{\underset{\sim}{2}}{\rho}(\vec{x}) \equiv \rho(\vec{x}) * \rho^-(\vec{x}) \qquad (1.3)$$

A useful method for visualizing convolution procedures is given by Hosemann and Bagchi[32b] and Papoulis[49a]. The two possible ways between $\rho(\vec{x})$ and $I(\vec{b})$ in (1.1) are due to the convolution theorem[32a, 49b]:

Given two arbitrary functions $a(\vec{x})$ and $g(\vec{x})$ and $A(\vec{b}) = Fa(\vec{x})$, $G(\vec{b}) = Fg(\vec{x})$ then:

$$F[a(\vec{x}) * g(\vec{x})] = A(\vec{b}) \cdot G(\vec{b}).$$

This is identical with (1.1) if we substitute:

$$a(\vec{x}) = \rho(\vec{x}),\ g(\vec{x}) = \rho^-(\vec{x})$$

$I(\vec{b})$ gives the scattering probability of incoming X-ray quanta for a given scattering vector $\vec{b}$. The actual differential scattering cross section or intensity $\sigma'(\Omega)d\Omega$ for elastic scattering can be calculated by the well known Ewald construction[32]. Within the small-angle region $1/\lambda \gg |\vec{b}|$ (in our case for Cu K_α radiation, $\lambda = 1{,}54$ Å) the Ewald sphere can be approximated as a plane containing the origin of the reciprocal space and extending perpendicular to the unit vector $\hat{b}_x$ (primary beam marks $\hat{b}_x$-axis of reciprocal space). Using polar coordinates b_r and b_φ within the Ewald-plane: $(b_x = 0, b_r, b_\varphi)$ (s. Fig. 4) one obtains from the Fourier-Optics approximation:

$$\sigma'(\Omega)\,d\Omega = I(b_x = 0, b_r = \vartheta/\lambda, b_\varphi = \varphi) \cdot f_e^2 \cdot f_\theta^2 \, d\Omega \tag{1.4}$$

where f_e^2 is the scattering factor of the single electron and f_θ^2 the polarization factor[32a]. The following relations can be deduced from the geometrical relationships of Fig. 3:

$$d\Omega = \sin\vartheta\, d\vartheta\, d\varphi \cong \vartheta\, d\vartheta\, d\varphi; \; dF_E = b_r\, db_r\, db_\varphi = \vartheta/\lambda^2 \, d\vartheta\, d\varphi$$

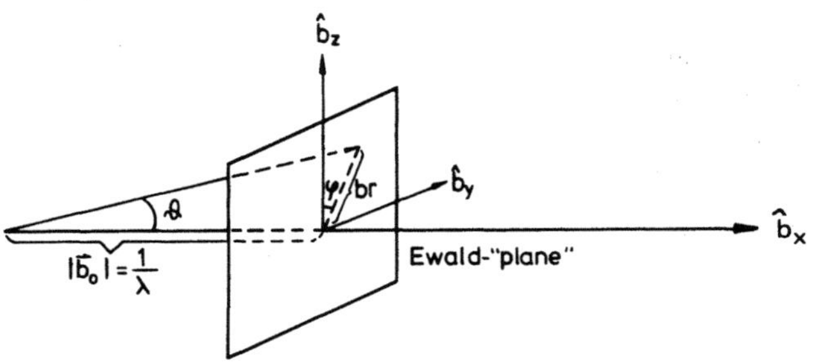

Fig. 3. Scattering geometry and "Ewald plane" in reciprocal space (Fourier space); $\vec{b}_0$ is the reciprocal vector of the primary beam

The experimental intensity I_R is measured on a registration plane at a distance PF from the sample (a film for instance), as shown in Fig. 4. If we use the polar coordinates r, φ within this registration plane, we get:

$$\begin{aligned} dF_R &= r\,dr\,d\varphi = \vartheta \cdot PF^2 \cdot d\vartheta\, d\varphi \\ I_R\,(r = \vartheta \cdot PF, \varphi)\, dF_R &= \sigma'(\vartheta,\varphi) \cdot I_{prim}\, d\Omega \end{aligned} \tag{1.5}$$

I_{prim} is the intensity of the primary beam.

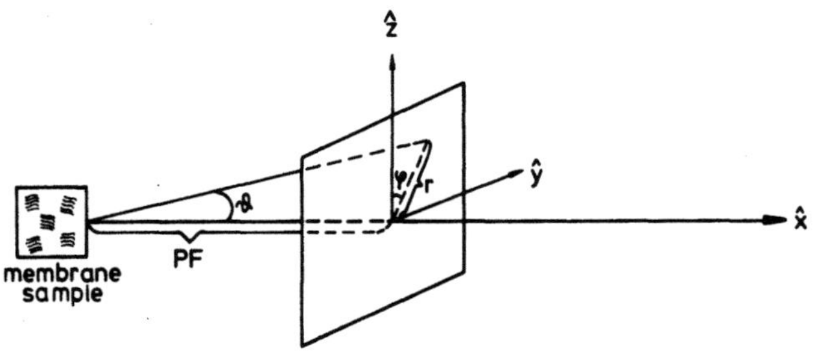

Fig. 4. Scattering geometry and registration plane in real space

Eliminating $\sigma'(\Omega)d\Omega$ from (1.4) and (1.5) we get:

$$I_R(r, \varphi) = I\left(b_x = 0, b_r = \frac{r}{PF \cdot \lambda}, b_\varphi = \varphi\right) \cdot f_e^2 \cdot f_\theta^2 \cdot \frac{I_{prim}}{PF^2} \tag{1.6}$$

In small-angle X-ray scattering, f_θ^2 can be taken as constant and thus the registration plane intensity is up to a constant equal to a section of the Ewald plane with the function $I(\vec{b})$. In the following we will calculate all relations in reciprocal space and omit the step to the registration intensity. It should be mentioned that Eq. (1.6) is valid only at distances PF from the membrane sample, where the details of the registration intensity I_R can be assumed to be large compared with the dimensions of the exposed sample volume.

1.3 Projection Theorem

Given two arbitrary functions $a(\vec{x}), g(\vec{x})$ in real space and the corresponding transforms in Fourier space: $A(\vec{b}), G(\vec{b})$, then the following relations hold due to the convolution theorem:

$$\begin{array}{ccccc} H(\vec{b}) & = & A(\vec{b}) & \cdot & G(\vec{b}) \\ F\uparrow & & F\uparrow & & F\uparrow \\ h(\vec{x}) & = & a(\vec{x}) & * & g(\vec{x}) \end{array}$$

If $\{\vec{b}_U / \vec{b}_U \in U\}$ is the set of all vectors, starting from the origin $\vec{b} = 0$ and terminating on any plane or straight line U passing through the origin and if $P_U a(\vec{x})$ is the orthogonal projection of $a(\vec{x})$ onto the corresponding plane or straight line U in real space, then the following relations hold:

$$\begin{array}{ccc} H(\vec{b}_U) & = & A(\vec{b}_U) \cdot G(\vec{b}_U) \\ F\uparrow & & F\uparrow \qquad F\uparrow \\ P_U h(\vec{x}) & = & P_U a(\vec{x}) * P_U g(\vec{x}) \end{array}$$

The proof is given for the simple case, that U is the b_z-axis of the Fourier space:

If we write down the Fourier transform in Cartesian coordinates:

$$H(b_x, b_y, b_z) = \int\int\int dx\, dy\, dz\, h(x, y, z)\, e^{-2\pi i (b_x \cdot x + b_y \cdot y + b_z \cdot z)}$$

we get the following expression by inserting the vectors of the b_z-axis of the Fourier space:

$$\begin{aligned} H(b_U) = H(0, 0, b_z) &= \int d_z \left\{\int dx \int dy\, h(x, y, z)\right\} e^{-2\pi i b_z \cdot z} \\ &= F\{P_z h(\vec{x})\} \end{aligned}$$

In the same way:

$$A(b_z) = F\,\{P_z a(\vec{x})\};\; G(b_z) = F\,\{P_z g(\vec{x})\} \tag{1.7}$$

Per definition:

$$H(0, 0, b_z) = A(0, 0, b_z) \cdot G(0, 0, b_z)$$

Inserting (1.7) and using the convolution theorem:

$$F P_z\, h(\vec{x}) = F\,\{P_z\, a(\vec{x}) * P_z\, g(\vec{x})\}$$

Applying F^{-1}:

$$P_z\, h(\vec{x}) = P_z\, a(\vec{x}) * P_z\, g(\vec{x}) \quad \blacksquare$$

The projection theorem assumes its most important form for our application, when we substitute especially:

$$a(\vec{x}) = \rho(\vec{x}) \qquad g(\vec{x}) = \rho^-(\vec{x})$$

$$\begin{array}{ccc} I(0, 0, b_z) & \xleftarrow{\;|\,|^2\;} & F(0, 0, b_z) \\ F\uparrow & & F\uparrow \\ P_z\, Q(\vec{x}) & \xleftarrow[\sim]{\;2\;} & P_z\, \rho(\vec{x}) \end{array}$$

The significance of this theorem follows from the special properties of membrane stacks:

1) Membranes are often flattened to form almost plane discs.
2) Membranes frequently have a lattice order within their surfaces (lateral lattice). Adjacent membranes however show no orientational correlation of their lateral lattices.

This illustrates the significance of projecting a membrane stack on its stacking axis (z-axis): It is not possible to obtain three-dimensional information on the electron density of the membranes. The projection of the membrane stack density on its stacking axis and some information on the lateral lattice is the best we can usually obtain from small-angle X-ray experiments.

1.4 Random Variables

We start with the concept that when building a membrane sample from its molecular components, numerous complicated processes take place, about which we have no detailed information, but which can be described by probability and random variables (r.v.), i.e. by statistical methods[50]. When examining membrane samples in electron microscope photographs we observe the following statistical properties:

- the axis of the membrane stacks shows an orientation probability distribution,
- the distance between the centres of gravity of the stacks is distributed randomly,
- the number of vesicles contained in a stack shows a probability distribution,
- the intermembrane distance within one stack shows a probability distribution,
- the surface of membranes is statistically uneven,
- particles within the membrane show probabilistic distributions.

To each of these *random variables* of the membrane sample belongs a density function or frequency distribution[50]. If a new sample is made under the same experimental conditions, the same random variables are at work and the membrane stackings represent another set of repeated trials[50] or sample of these variables. When calculating diffraction intensities, we often have to evaluate the mean of these values, which a special random variable (r.v.) assumes in our membrane sample (for instance intermembrane distances, number of vesicles of the stack).

The *law of large numbers*[50] enables us to substitute in all these cases the mean of infinitely many trials or the expectation value of the r.v.[50]. This greatly simplifies the results and reduces the number of unknown variables.

1.5 Incoherent Superposition, *Flower Bunch Statistics*

Membrane stacks show random distances of their centres of gravity when viewed by electron microscopy. The stacking axes show a frequency maximum in a particular direction called the *z-axis* which is determined by the preparation process. Shifting all stacking axes parallel to a common origin, one obtains a picture as in Fig. 5.:

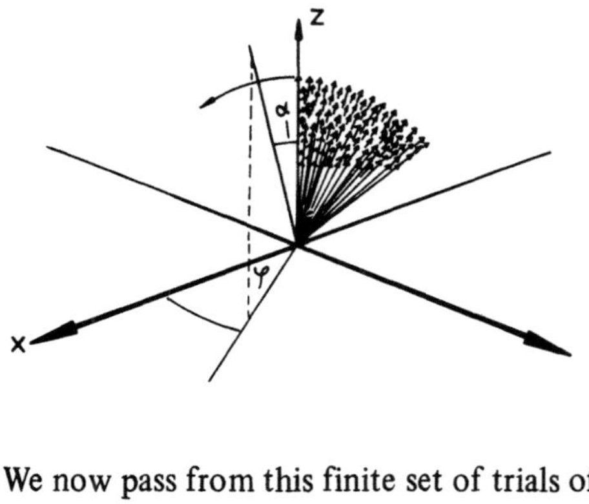

Fig. 5. Orientation distribution of stacking axes of a membrane sample (Flower bunch)

We now pass from this finite set of trials of the particular membrane sample to the infinite set of trials, whose density distribution is assumed to be independent of the azimuthal angle φ and can be written $h(\alpha)$ sin (α). In the literature this function is frequently called *mosaic spread,* in this paper we prefer the more illustrative name *flower bunch statistics.* The set of all stacks of our particular membrane sample having an axis in the solid angle element $(\alpha, d\alpha, \varphi, d\varphi)$ can be numbered by the index s. Owing to the random centre-to-centre distances, the scattering intensity $J_{\alpha,\varphi}$ of this set of stacks can be written as incoherent superposition of the intensities I_s of its individual elements:

$$J_{\alpha,\varphi} = \sum_s \mathrm{I}_s = \sum_s |F\boldsymbol{\rho}_s|^2 = \sum_s F\overset{2}{\tilde{\boldsymbol{\rho}}}_s = \sum_s F\mathbf{Q}_s \tag{1.8}$$

Taking the Fourier transform:

$$F^{-1}J_{\alpha,\varphi} \equiv \mathbf{Q}_{\alpha,\varphi} = \sum_s \mathbf{Q}_s = \sum_s \overset{2}{\tilde{\boldsymbol{\rho}}}_s$$

Neglecting a constant factor, $J_{\alpha,\varphi}$ is the mean of the sample trials: $\langle \mathrm{I}_s \rangle_s$. We now use the law of great numbers to substitute the infinite trial mean $\mathrm{I}^*_{\mathrm{ST}}$ for this sample mean:

$$J_{\alpha,\varphi} = h(\alpha)\sin\alpha\, \mathrm{d}\alpha \frac{\mathrm{d}\varphi}{2\pi}\, \mathrm{I}^*_{\mathrm{ST}}; \quad \mathrm{I}^*_{\mathrm{ST}} = F\langle \overset{2}{\tilde{\boldsymbol{\rho}}}_s \rangle \equiv F\mathbf{Q}^*_{\mathrm{ST}}$$

We assume that all r.v. of the set of stacks within $(\alpha, \mathrm{d}\alpha, \varphi, \mathrm{d}\varphi)$ are independent of α and φ and that $\mathrm{I}^*_{\mathrm{ST}}$, $\mathbf{Q}^*_{\mathrm{ST}}$ are cylindrical symmetrical with respect to their own stacking axis. We get the total intensity diffracted by all membrane stacks by adding all $(\alpha, \mathrm{d}\alpha, \varphi, \mathrm{d}\varphi)$ contributions. This becomes easy if we use spherical coordinates $\mathrm{b_r}, \vartheta$; r, ϑ to describe $\mathrm{I}^*_{\mathrm{ST}}$; $\mathbf{Q}^*_{\mathrm{ST}}$ with regard to their own polar axes (a detailed description is given in[51)]). We will use only the diffraction intensity along the $\mathrm{b_z}$-axis, which is a superposition of all $J_{\alpha,\varphi}$ contributions:

$$\mathrm{I}^*_{\mathrm{exp}}(\mathrm{b_z}) \equiv \mathrm{I}(\mathrm{b_z}, \vartheta = 0) \tag{1.9}$$

$$\mathrm{I}^*_{\mathrm{exp}}(\mathrm{b_z}) = \int_0^{\pi} h(\alpha)\sin\alpha\, \mathrm{I}^*_{\mathrm{ST}}(\mathrm{b_z}, \alpha)\, \mathrm{d}\alpha$$

1.6 Separation of Lateral Structure from Layer Structure

Usually the membrane surfaces show undulations with typical dimensions as shown in Fig. 6a. The simplest way to describe such a wavy membrane is to use a coordinate system x′, y′, z′ with coordinate surfaces parallel to the membrane surface, as shown in Fig. 6b.

The electron density of the membrane $\boldsymbol{\rho}_{\mathrm{M}}$ (the bold types indicate a three dimensional density) for the case of a *pure layer structure* can be written:

$$\boldsymbol{\rho}_M = \rho_M(\mathrm{z}') \cdot 1(\mathrm{x}') \cdot 1(\mathrm{y}')$$

Fig. 6a. Schematic drawing of typical membrane undulations

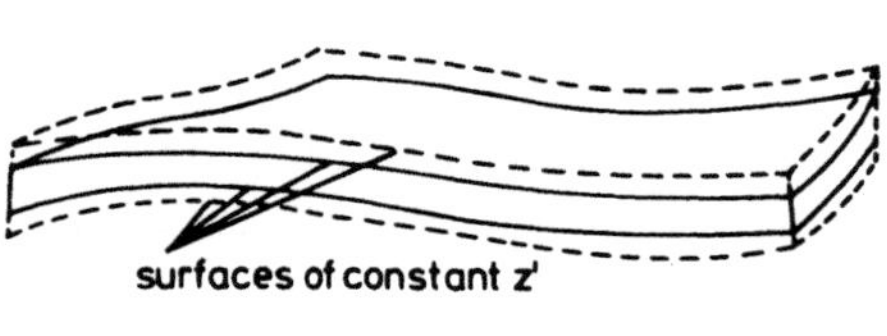

Fig. 6b. An undulated membrane with its curvilinear coordinate system

The layer structure is an oversimplified model, since the membranes also have a lateral structure. However the real membrane density can be separated into a pure layer structure and a lateral density $\Delta\rho_M$:

$$\rho_M(x', y', z') = \rho_M(z') \cdot 1(x') \cdot 1(y') + \Delta\rho_M(x', y', z') \tag{1.10}$$

ρ_M is evaluated by projecting the electron density along the z' = const. surfaces onto any x' = const. y' = const. axis. We assume further that all membranes have the same layer profile ρ_M.

Given a particular curvilinear system x', y', z', we can define the subset of all membranes (of an infinite set), whose wavy surface is parallel to the sheets z' = const. It seems reasonable to assume that the lateral structure does not depend on the particular surface form of the membranes. We therefore conclude that the layer structure

$$\rho_M(z') \cdot 1(x') \cdot 1(y')$$

is identical with the mean electron density of the set selected above. In the same way the lateral densities of the set have a vanishing mean. To continue we again consider the infinite set of membrane stacks with the same stacking axis, which was already used to calculate I^*_{ST} in (1.8). Each membrane stack can be replaced by an imaginary stack containing the layer structure $\rho_M(z') \cdot 1(x') \cdot 1(y')$ of each membrane and an imaginary stack containing the lateral structure $\Delta\rho_M$ of each membrane. The scattering intensity I^*_{ST} of the membrane stacks can be proved as the sum of the scattering intensity of the layer stacks I_{ST} and of the lateral-structure stacks, as is shown in 5.1:

$$I^*_{ST} = I_{ST} + \langle n(s)\rangle \cdot F\langle \overset{2}{\widetilde{\Delta\rho^r_s}} + \overset{2}{\widetilde{\Delta\rho^e_s}}\rangle \tag{1.11}$$

Both kinds of stacks now show the same stacking axis distribution $h(\alpha)\sin(\alpha)$. The layer stacks therefore scatter into a cone whose axis is identical with the b_z-axis, while the lateral structure scatters with the same cone-angle along the $b_x - b_y$ plane (s. Fig. 7).

Except from layer intensity and lateral intensity we have an additional background intensity, caused by disordered macromolecules in solution, parts of membranes and micelles (s. Fig. 7).

The principal aim of this paper is to calculate the layer profile ρ_M and the information required is the layer intensity on the b_z-axis:

$$I_{exp}(b_z) = \int_0^\pi h(\alpha)\sin(\alpha)\, I_{ST}(b_z, \alpha)\, d\alpha \tag{1.12}$$

$I_{exp}(b_z)$ can be separated from the measured superposition of I_{ST} and the background intensity in the b_z direction by subtracting the intensity measured along an axis a (s. Fig. 7) at an angle where the background intensity predominates over the other two contributions from the intensity measured along the b_z-axis[40,41,43,48]

In the following we shall assume the layer intensity on the b_z-axis $I_{exp}(b_z)$ is known except for the immediate neighbourhood of the origin $b_z = 0$.

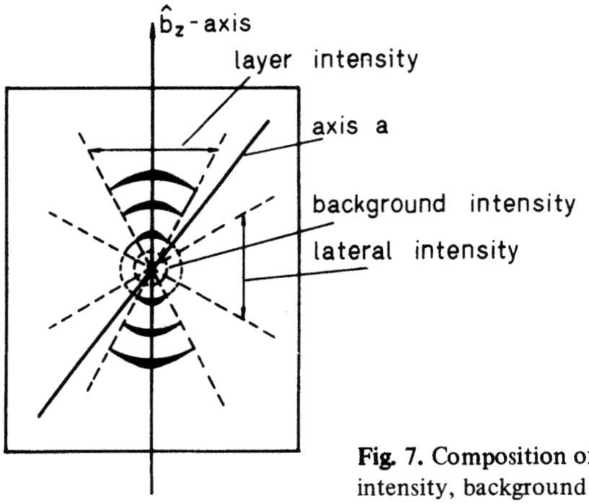

Fig. 7. Composition of the registration plane intensity by layer intensity, background intensity and lateral intensity (cf. Fig. 2)

1.7 Plan for Further Chapters

The information contained in the scattering intensity or Q-function and the best method for obtaining it, i.e. calculating the layer profile ρ_M, depends strongly on the frequency distributions of the random variables (r.v.) governing the membrane sample. The less random they are, the easier the evaluation of the layer profile usually is.

An ensemble of small stacks, consisting only of a few, identical, equidistant and plane vesicles (pairs of two mutually symmetrical membranes) appeared to be a model that was a sufficient basis in the past for evaluating ρ_M from small-angle X-ray diagrams. In Chap. 2 we will show how to derive the Q-function of this model and evaluate the membrane layer profile from it. Later we shall discuss its shortcomings and the requirements for more realistic assumptions about the random variables (r.v.) of the membrane stacks. Chap. 3 will introduce the resulting disorder of the intermembrane distances and Chap. 4 statistical undulations of the membrane surface.

2 Small Crystalline Stacks of Plane Layers

2.1 The Assumptions and the Q-Function

Particularly assumed in this model:

(2.1) The stackings consist of only a few plane vesicles (about 1–5)

(2.2) In each vesicle two membranes are mirror symmetrically arranged, i.e. the electron density ρ_0 of the whole vesicle possesses a centre of symmetry.

(2.3) Centre-to-centre distances of membranes within one vesicle are all identical. Centre-to-centre distances of adjacent membranes of two different vesicles are all identical.

(2.4) The dimensions of the membranes parallel to the surface are large compared with the height of the stacks.

The distribution of the stacking axes, flower bunch statistics or mosaic spread, and the number of elementary cells (vesicles) within the stacks are the only r.v. that are allowed to be really statistically variable in this model.

Due to (2.1), (2.4) we can assume I_{ST} to be concentrated into a δ-cylinder around the b_z-axis (δ designates the delta function):

$$I_{ST}(b_r, \alpha) = \frac{\delta(b_R)}{\pi b_R} \cdot I_z(b_z) \text{ with } I_z(b_z) \equiv I_{ST}(0, 0, b_z) \tag{2.5}$$

Here we have used spherical coordinates on the left side and cylindricals on the right side with the conversion formula:

$$\begin{aligned} b_z &= b_r \cos\alpha \\ b_R &= b_r \sin\alpha \\ \varphi &= \varphi \end{aligned} \tag{2.6}$$

Substituting (2.5) and (2.6) into (1.12):

$$I_{exp}(b_r) = \int_0^{\pi} h(\alpha) \sin(\alpha)\, I_z(b_r \cos\alpha) \cdot \frac{\delta(b_r \sin\alpha)}{\pi b_r \sin\alpha}\, d\alpha$$

Using the well known relation[49c)]:

$$\delta(f(x)) = \sum_i \frac{\delta(x - x_{oi})}{|f'(x_{oi})|} \quad \text{where } x_{oi} \text{ are the zeros of } f(x) \tag{2.7}$$

we obtain:

$$b_r^2\, I_{exp}(b_r) = \text{const.} \cdot I_{ST}(0, 0, b_r) \tag{2.8}$$

The distribution of the stacking axes is corrected for by simply multiplying the experimental layer intensity $I_{exp}(b_z)$ with the well known *Lorentz factor (platelet factor)* to obtain the scattering intensity I_{ST} of ideally oriented stacks with parallel axes. Using the projection theorem we obtain from (1.8):

$$\begin{aligned} F^{-1}\{I_{ST}(0, 0, b_z)\} &= P_z\, Q_{ST} = \langle P_z \overset{2}{\tilde{\rho}_s} \rangle_s = \langle \overset{2}{\widetilde{P_z \rho_s}} \rangle_s \\ &\equiv \langle \overset{2}{\tilde{\rho}_s} \rangle_s \equiv Q_{ST} \end{aligned} \tag{2.9}$$

By $n(s)$ we denote the number of vesicles, contained in the stack with index s.

Using the assumptions (2.2), (2.3), (2.4) we can write:

$$\rho_s(z) = \sum_{i=1}^{n(s)} \rho_0(z) * \delta(z - i \cdot L) \tag{2.10}$$

where we have summarized both profiles of the vesicle into:

$$\rho_0(z) \equiv \rho_M(z) + \rho_{\bar{M}}(z) * \delta(z - L/2) \tag{2.11}$$

Starting with (2.10) we now calculate the Q_{ST}-function:

$$\overset{2}{\tilde{\rho}_s} = \{\sum_{i=1}^{n(s)} \rho_0 * \delta(z - i \cdot L)\} * \{\sum_{k=1}^{n(s)} \rho_{\bar{0}} * \delta(z + k \cdot L)\}$$

$$= \sum_{i}^{n(s)} \sum_{k} \overset{2}{\tilde{\rho}_0} * \delta(x - (i - k) \cdot L)$$

Using a new index j = i − k for the series we obtain (s. Fig. 9)

$$\overset{2}{\tilde{\rho}_s} = \sum_{j=-(n(s)-1)}^{n(s)-1} \sum_{i=\mathrm{Max}(1,\, j+1)}^{\mathrm{Min}(n(s),\, n(s)+j)} \overset{2}{\tilde{\rho}_0} * \delta(z - j \cdot L); \quad (n(s) - |j|)_p \equiv \begin{cases} n(s) - |j| & |j| \leq n(s) \\ 0 & |j| > n(s) \end{cases}$$

$$= \sum_{j=-\infty}^{\infty} (n(s) - |j|)_p \cdot \overset{2}{\tilde{\rho}_0} * \delta(z - j \cdot L)$$

We define:

$$Q_0 \equiv \overset{2}{\tilde{\rho}_0}$$

From (2.9):

$$Q_{ST} = \langle \overset{2}{\tilde{\rho}_s} \rangle_s = \sum_{j=-\infty}^{+\infty} \langle (n(s) - |j|)_p \rangle \cdot Q_0 * \delta(z - j \cdot L) \tag{2.12}$$

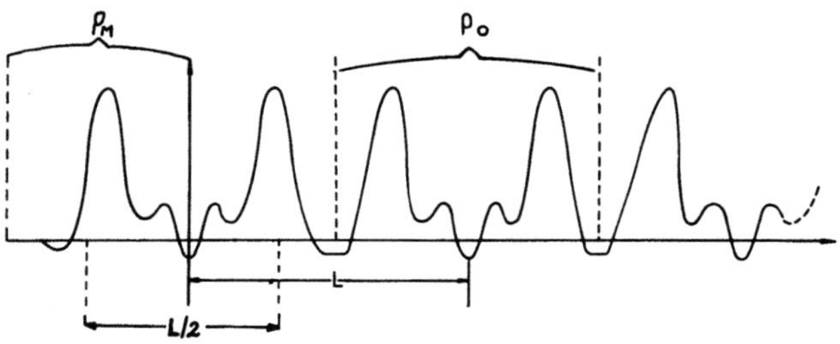

Fig. 8. Section of a layer profile of a small crystalline membrane stack

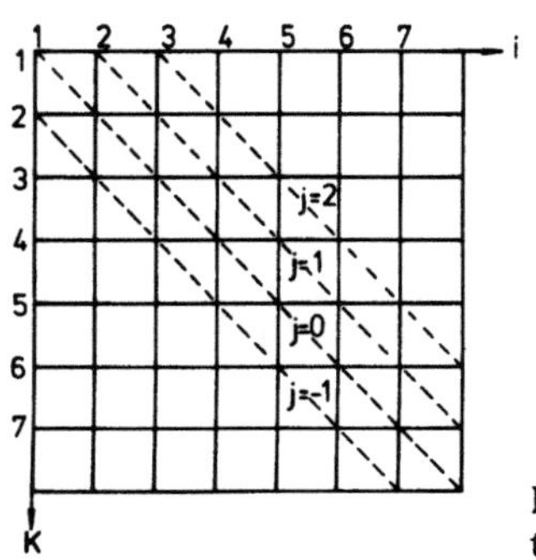

Fig. 9. Schematic representation substitution of summation over i, j for i, k

s is again indexing an infinite set of stacks, all ruled by the r.v. defined in (2.1–2.4) which we assume to hold in our actual membrane preparation. As can be seen from (2.12) the Q_{ST} function consists of a series of symmetric functions $\tilde{\rho}_0^2 = Q_0$ centred on the *mirror points* j · L and extending maximally between the two adjacent points to the left (j – 1) · L and to the right (j + 1) · L.

A schematic diagram of Q_{ST} between the j = 0 and j = 1 mirror points is given in Fig. 10.:

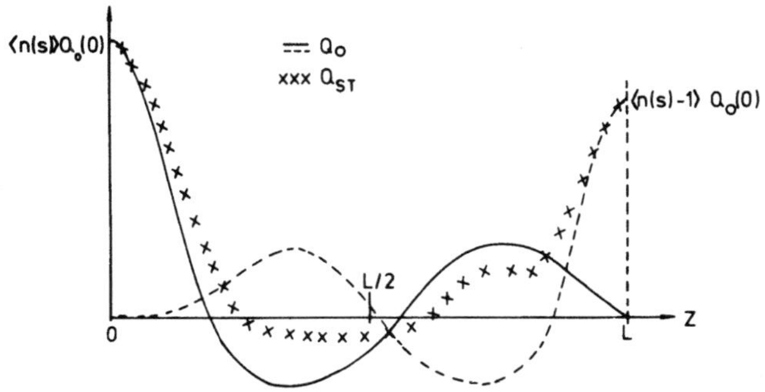

Fig. 10. Schematic representation composition of Q_{ST} within the first Bragg-period as a superposition of two mirror symmetrical arranged Q_0-functions with different weight

Abbreviation: $g_0 \equiv \langle n(s) \rangle$, $g_1 = \langle n(s) - 1 \rangle$

2.2 Isolation of the Q_0-Function

The objective of our theory is to evaluate the vesicle profile ρ_0 from $I_{exp}(b_z)$. The first step towards this goal was to obtain Q_{ST} from (2.8) and (2.9). The next step is to solve Q_{ST} for Q_0. This aim can be reached either by "mirror symmetrical subtraction"[36, 39] or by the following procedure for which it will be useful to introduce the following definitions:

$$HQ_{ST}(z) \equiv \begin{cases} Q_{ST}(z + L/2) & -L/2 \leqslant z \leqslant L/2 \\ 0 & \text{otherwise} \end{cases}$$
$$HQ_0(z) \equiv \begin{cases} Q_0(z + L/2) & -L/2 \leqslant z \leqslant L/2 \\ 0 & \text{otherwise} \end{cases} \tag{2.13}$$

Symmetric part of HQ_{ST}:

$$HQ_{ST}^s \equiv {}^1\!/_2\,(HQ_{ST} + HQ_{ST}^-)$$

Antisymmetric part of HQ_{ST}: $HQ_{ST}^a \equiv {}^1\!/_2\,(HQ_{ST} - HQ_{ST}^-)$
HQ_0^s and HQ_0^a are defined in a similar way.

Regarding Fig. 10 we find:

$$HQ_{ST} = HQ_{ST}^s + HQ_{ST}^a = (g_0 + g_1) \cdot HQ_0^s + (g_0 - g_1) \cdot HQ_0^a$$

We solve for HQ_0:

$$HQ_0 = HQ_0^s + HQ_0^a = \frac{HQ_{ST}^s}{g_0 + g_1} + \frac{HQ_{ST}^a}{g_0 - g_1} = \frac{HQ_{ST} + HQ_{\overline{ST}}}{2(g_0 + g_1)} + \frac{HQ_{ST} - HQ_{\overline{ST}}}{2(g_0 - g_1)} \tag{2.14}$$

Since HQ_0 contains all information on Q_0, we are now faced with the problem of solving the integral equation:

$$Q_0(z) = \int dz' \, \rho_0(z') \, \rho_0(z' - z) = \int dz' \, \rho_0(z') \, \rho_0(z' + z) \tag{2.15}$$

for the unknown, symmetric profile ρ_0.

Formulated in reciprocal space, we have to solve

$$I_0(b) = (F_0(b))^2$$

for $F_0(b)$, with a real function $F_0(b)$. The following three procedures are used to solve this problem. They all give the same two solutions, which differ by a sign:

1) Reconstruction method (for a detailed description see[13)]): we take the square-root of $I_0 = \mathcal{F} Q_0$ to obtain the modulus of the structure amplitude $|F_0(b)|$. On the other hand we can reconstruct the latter function with the help of Shannon's theorem[52)] from only a set of equidistant points $|F_0(k/L)|$ with L being equal (or larger than) the extension of the electron density:

$$|F_0(b)| = \left| \sum_k x_k \, |F_0(k/L)| \cdot \frac{\sin(\pi L b - \pi k)}{\pi L b - \pi k} \right|, \quad x_k = \pm 1$$

But this equation holds for the true phase combination and another one with + and − interchanged only. By testing all possible phase combinations, these two combinations can be found. The two possible solutions of (2.15) are given by:

$$\rho_0(z) = \sum_k x_k \, |F_0(k/L)| \exp\left(2\pi i \frac{k}{L} z\right)$$

2) Hosemann's deconvolution procedure[32c)].
3) Pape's Fourier analytical deconvolution procedure[33)].

Pape's method played an important role in the development of the more complicated procedures to be described later. We therefore rederive it here in a more suitable form, than that given in the original publication[33)].

2.3 Pape's Fourier Analytical Deconvolution

The following integral equation must be solved for the real, symmetrical function ρ_0.

$$Q_0(z) = \int dy \, \rho_0(y) \, \rho_0(z + y) = \int dy \, \rho_0(y) \, \rho_0(y - z) \tag{2.16}$$

Developing ρ_0 as well as Q_0 into a Fourier series and inserting it into (2.16) reveals a simple relation between both sets of Fourier coefficients that can be solved except for one sign for the coefficients of ρ_0. The vesicle profile of width L is mapped into the interval $-\pi \leqslant z \leqslant \pi$, then the Fourier series for ρ_0 becomes:

$$\underset{-\pi\leqslant z\leqslant\pi}{\rho_0(z)} = \sum_{k=-NC}^{NC} c_k\, e^{ikz} \tag{2.17}$$

and since ρ_0 is real: $c_k = c^*_{-k}$ and since ρ_0 is symmetric: $c_k = c_{-k}$. Substituting (2.17) in (2.16) one has to take into consideration the limits of integration, since the Fourier-series (2.17) resembles ρ_0 only in the interval $-\pi \leqslant z \leqslant \pi$.

$$\underset{z\geqslant 0}{Q_0(z)} = \int_{-\pi+z}^{+\pi} \left\{ \sum_k c_k\, e^{iky} \sum_l c_l\, e^{il(y-z)} \right\} dy$$

After a straightforward calculation (see 5.2.1):

$$\underset{z\geqslant 0}{Q_0(z)} = \sum_{\substack{k,\,l\\ k\neq l}}\sum c_k\, c_l\, \frac{i(-1)^{k+l}}{k+l}\,[e^{ikz} - e^{-ilz}] + \sum_k c_k\, c_{-k}\, e^{ikz}(2\pi - z) \tag{2.18}$$

We now calculate the Fourier-coefficients I_n of a series, representing $HQ_0(z-\pi)$ within its period $(0, 2\pi)$

$$I_n \equiv \int_0^{2\pi} Q_0(z)\, e^{-inz}\, dz \tag{2.19}$$

Substituting (2.18) we obtain (5.2.2):

$$I_n = \sum_{k\neq -n} i\,\frac{2\pi(-1)^{n+k}}{n+k}\, c_n\, c_k + \sum_{k\neq n} i\,\frac{2\pi(-1)^{k-n}}{n-k}\, c_k\, c_{-n}$$

$$+ \sum_{k\neq n} i\,\frac{2\pi(-1)^{n+k}}{k-n}\, c_{-k}\, c_n + 2\pi\, c_n\, c_{-n}$$

If we now substitute $c_k = c_{-k}$ and split into real and imaginary parts, we get:

$$\mathbf{Re}\, I_n = 2\pi^2\, c_n^2 \tag{2.20}$$

$$\mathbf{Im}\, I_n = \sum_{n\neq k} c_n\, c_k\, \frac{4\pi(-1)^{n+k}}{n-k} + \sum_{k\neq n} c_k^2\, \frac{2\pi}{k-n} \tag{2.21}$$

The numbers I_n can be calculated from the known Q_0-function. It is possible to solve for the unknown Fourier coefficients in the following way: From (2.20):

$$|c_n| = \frac{1}{\pi}\sqrt{\frac{\mathbf{Re}\, I_n}{2}} \tag{2.22}$$

We now insert $c_n = |c_n| \cdot x_n$ into (2.21) with the moduli from (2.22) and the unknown phases $x_n = \pm 1$. After some rearrangements we obtain a linear, homogenous system of equations:

$$\sum_{k \neq n} \left[\frac{4\pi\, |c_n|\, |c_k|\, (-1)^{n+k}}{n-k} \right] x_k + \left\{ \sum_{k \neq n} \frac{2\pi |c_k|^2}{k-n} - \mathbf{Im}\, I_n \right\} x_n = 0 \qquad (2.23)$$

There are well known methods of finding all possible solutions to these equations. The special sign-vectors x_n are found when the condition, that all its coefficients have a modulus of 1, is satisfied.

Remarks: The above procedure can be modified in the case of an antisymmetric function $\rho_0(-x) = -\rho_0(x)$ and also to multidimensional convolution squares of functions possessing inversion symmetry or antisymmetry.

Mahler[28] could rederive the above equations by using the relation $I_0 = (F\rho_0)^2$ instead of the convolution integral (2.16).

The system of equations, whose solution provides the unknown phases, mainly depends on the integrals:

$$\mathbf{Im}\, I_\eta = -\int_0^{2\pi} Q_0(z) \sin(\eta z)\, dz = -i \int_0^{2\pi} HQ_0^a(z-\pi)\, e^{-i\eta z} dz$$

$$= \frac{-i}{2(\langle n \rangle - \langle n-1 \rangle)} \int_0^{2\pi} (HQ_{ST} - HQ_{\overline{ST}})\, e^{-i\eta z}\, dz$$

Stackings with many elementary cells have $HQ_{ST} \cong HQ_{\overline{ST}}$ and $\langle n \rangle \cong \langle n-1 \rangle$. Evidently the evaluation of the phases becomes less sensitive to experimental errors as the number of vesicles in the stacks decreases. From the point of view of the reciprocal space we have broad reflections, whose centre is slightly displaced from the reciprocal lattice points in this case. This displacement can be used to determine the phases.

The methods above were recently applied to the scattering intensity of artificial fatty acid bilayers[28]. We have already mentioned that ρ_0 can also be obtained by taking the square root of $I_0(b)$. Knowing I_0 should therefore be equivalent to knowing the integrals $\mathbf{Re}\, I_n$ and $\mathbf{I_m}\, I_n$.

Though this is already obvious from the derivation of these numbers, an other interesting proof found by Mahler[28] should be mentioned. Knowing $I_0(b)$ is equivalent, from Shannon's theorem, to certain equidistant points of I_0. In our case, with $FI_0 = Q_0$ extending from -2π to 2π, Shannon's theorem can be written:

$$I_0(b) = \sum_{n=-\infty}^{+\infty} I_0\left(\frac{n}{4\pi}\right) \frac{\sin(4\pi^2 b - n\pi)}{4\pi^2 b - n\pi}$$

Using stacks with many elementary cells, the scattering intensity only gives the values $I_0(2n/4\pi)$, which suffices for determining the moduli of the Fourier coefficients of ρ_0, but still lack the phases. Mahler derived the following relation for these *even-numbered* Shannon-points:

$$I_0\left(\frac{2n}{4\pi}\right) = \text{const. } \mathbf{Re}\, I_n$$

The remaining *odd* points $I_0\left(\frac{2n+1}{4\pi}\right)$ should therefore suffice for solving the phase problem except for a sign. Mahler finds a linear relationship between these points and the numbers $\mathbf{I}_m\, I_n$ in the following form:

$$\mathbf{I}_m\, I_n = \sum_k I_0\left(\frac{2k+1}{4\pi}\right)\frac{4}{2(k-n)+1}$$

The method of evaluating HQ_0 and the phase determination with Pape's procedure are very sensitive to the validity of the assumption (2.3), i.e. that there are no lattice distortions. In 5.3 we describe a computer experiment demonstrating this sensitivity for distance disorder in the Q_0-function. Biological membranes usually show a distribution of intermembrane distances. Therefore a more general model of the r.v. in membrane samples has to be used, which is given in Chap. 3.

3 Paracrystalline Plane-Layer Lattice

3.1 The Assumptions and the Q-Function

Usually biomembranes form large stacks of more than 10 unit cells (vesicles). Nevertheless the corresponding Q_{ST}-functions converge to zero after having reached a distance, which is small compared with the number of unit cells $\langle n(s)\rangle_s$ of the stacks times their mean period $\langle L\rangle$.

Furthermore, a smearing or smoothening of the Q_{ST}-function can be observed with increasing distance from the origin. Experimental examples of Q_{ST}-functions are given in Fig. 11.

We explain these characteristics as a consequence of a disorder of intermembrane distances, i.e. if we take pairs of vesicles out of the stack, all consisting of two vesicles being separated by a number m of vesicles lying between them, then with increasing m the std. dev. of distances between the centres of the two vesicles becomes larger proportional to $\sqrt{m}$ (*lattice distortions of the second kind*[32)]). As soon as this std. dev. has outpaced the wavelength Δ of the most prominent Fourier waves of the vesicle profile, the Q_{ST}-function will have converged to the correlation function of a homogenous electron density distribution, i.e. a correlation function of a electron density rectangle, the width of the latter being identical with the height of the stack. (Exactly it converges to a superposition of convolution squares of rectangles, since the length of the stack is a random variable). See Fig. 12 for a computer-generated Q_{ST}-function of this kind.

Since we assume long stacks, the *volume scattering*[32d)] of this rectangle of mean electron density of the stack is concentrated closely around $b_z = 0$ and cannot be measured. We may therefore assume that the mean electron density of the stacks is zero and the Q_{ST}-function converges to zero.

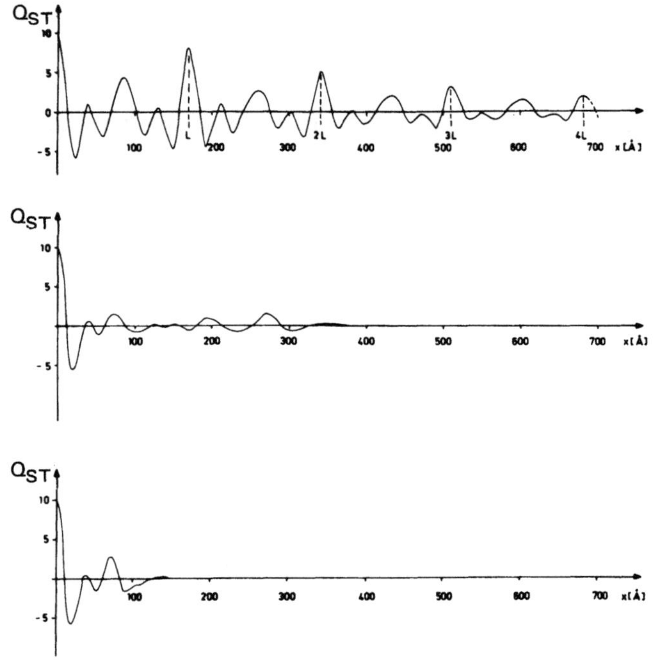

Fig. 11. Q_{ST}-functions of nerve-myelin prepared in three different ways showing increasing intermembrane distance disorder[43]

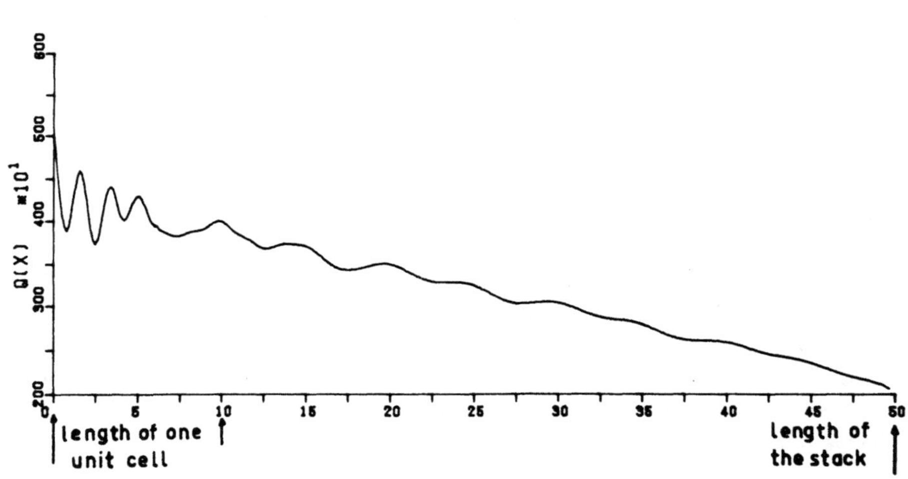

Fig. 12. Computer generated Q_{ST}-function, exhibiting convergence to a wedge by intermembrane distance disorder

Guided by this observation, we assume that the r.v. governing the stacks have the following properties:

(3.1) The membrane stacks consist of a large number of vesicles.

(3.2) Distances between adjacent membranes of adjacent vesicles show a probability distribution $H2(l_2)$. Distances between membranes within a vesicle (intravesicular distances) show a disorder characterized by a probability distribution $H1(l_1)$, which is uncorrelated with the H2 distribution.

(3.3) The dimensions of the membranes parallel to the surface are large compared with the *correlative height* of the stacks, i.e. the distance, when the Q_{ST}-function has converged to zero by the distance disorder.

(3.4) The surfaces of all membranes within a stack are plane and parallel to each other.

(3.5) The membrane layer profiles are all identical but are arranged in an image and mirror image sequence: $\rho_M, \rho_{\bar{M}}, \rho_M, \rho_{\bar{M}}, \ldots$

Repeating the derivation of eqs. (2.5)–(2.8), we again obtain $I_{ST}(0, 0, b_z)$ by simply multiplying $I_{exp}(b_z)$ by the Lorentz or platelet factor b_z^2:

$$b_z^2 I_{exp}(b_z) = \text{const.} \cdot I_{ST}(0, 0, b_z) \tag{3.6}$$

The following equation again holds:

$$F[I_{ST}(0, 0, b_z)] = P_z Q_{ST} = \langle \widetilde{P_z \rho_s}^2 \rangle = \langle \tilde{\rho}_s^2 \rangle_s = Q_{ST} \tag{3.7}$$

An arbitrary membrane stack ρ_s should visually consist of a series of identical membrane profiles arranged in an image and mirror image sequence, where the single profiles are separated by gaps of statistical width. These gaps are filled up by rectangles of constant electron density, as is shown in Fig. 13.

The Bragg period of such a structure is identical with the mean of the distances between two adjacent midpoints of intervesicular gaps.

Let us consider out of the set of all possible stacks the special stack $\rho_{s'}$, whose intra- (inter-) vesicular gaps all have a width equal to the mean width of all possible intra- (inter-) vesicular gaps. This special stack can easily be described by a *building block* ρ_M consisting of the membrane profile with the adhering halves of the mean inter- and intravesicular electron density rectangles (Fig. 14).

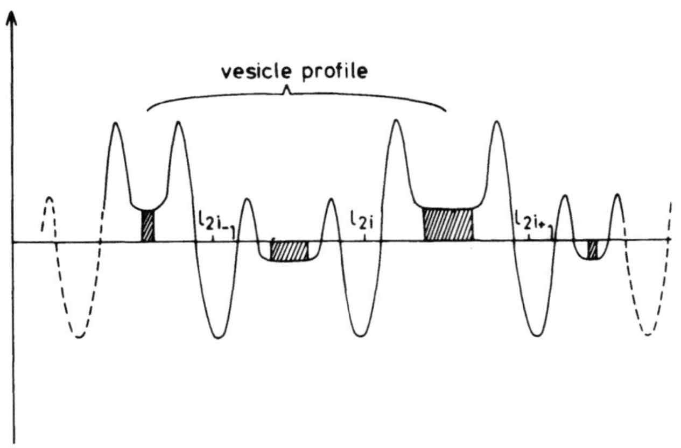

Fig. 13. Layer profile of a paracrystalline plane membrane stack showing the invariable membrane profiles separated by two types of gaps of statistical width $\langle l_1 \rangle = \langle l_{2i} - l_{2i-1} \rangle$, $\langle l_2 \rangle = \langle l_{2i+1} - l_{2i} \rangle$

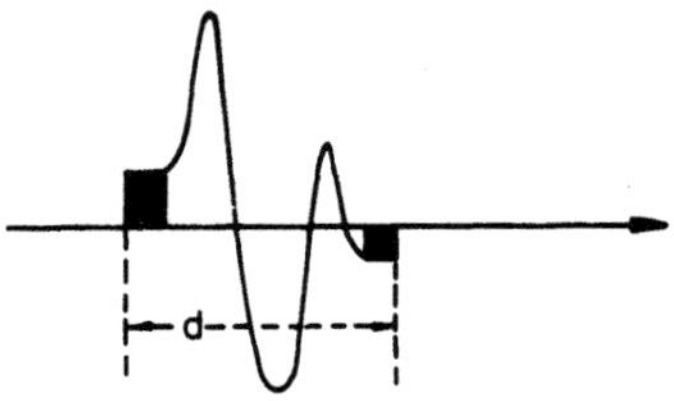

Fig. 14. Building block ρ_M consisting of the membrane profile and the adhering halfs of the mean inter- and intra-vesicular gaps

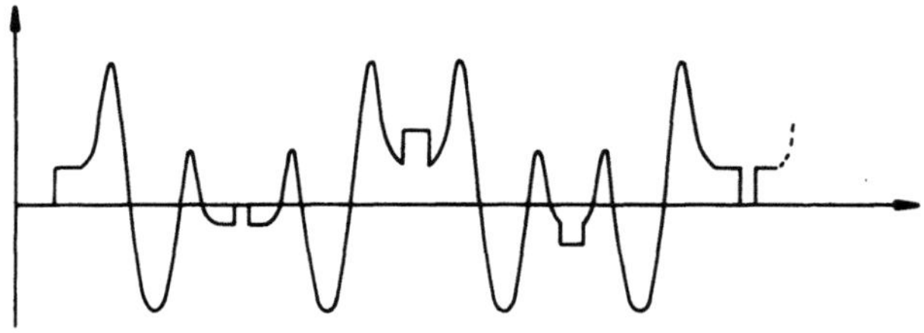

Fig. 15. Layer profile of a stack after substitution of ρ_M for each of the original membranes and gap-densities, showing artificial blanks and overlaps

We can simply write:

$$\rho_{s'} = \sum_{i=1}^{n(s')} \rho_M * \delta(x - (2i-1) \cdot d) + \rho_{\bar{M}} * \delta(x - 2i \cdot d)$$
$$d = \langle L \rangle / 2 \tag{3.8}$$

However if we apply the same building block ρ_M to any other stack ρ_s, i.e. if we replace each membrane profile and its adhering half gaps by the building block ρ_M defined above, we get an inadequate or incomplete representation of the stack exhibiting artificial overlaps and gaps as shown schematically in Fig. 15.

In order to obtain a consistent mathematical description of all stacks in terms of the building block ρ_M we have to add or subtract other rectangles, as given in (3.9), the first row of which is identical with the incomplete representation above. See Fig. 17 for an explanation of the additional terms.

$$\rho_s = \sum_{i=1}^{n(s)} \{\rho_M * \delta(x - l_{2i-1}) + \rho_{\bar{M}} * \delta(x - l_{2i})\}$$
$$+ \left\{ h_I \cdot \sigma_i^I \cdot Z(|l_{2i} - l_{2i-1} - d|) * \delta\left(x - \frac{l_{2i} + l_{2i-1}}{2}\right)\right\} \tag{3.9}$$
$$+ \left\{ h_E \cdot \sigma_i^E \cdot Z(|l_{2i+1} - l_{2i} - d|) * \delta\left(x - \frac{l_{2i} + l_{2i+1}}{2}\right)\right\}$$

with the definitions:

$$\sigma_i^I \equiv \mathrm{sgn}(l_{2i} - l_{2i-1} - d) = \begin{cases} +1 & \text{for } l_{2i} - l_{2i-1} > d \\ -1 & \text{for } l_{2i} - l_{2i-1} < d \end{cases}$$

$$\sigma_i^E \equiv \operatorname{sgn}(l_{2i+1} - l_{2i} - d) = \begin{cases} +1 & \text{for } l_{2i+1} - l_{2i} > d \\ -1 & \text{for } l_{2i+1} - l_{2i} < d \end{cases}$$

$$Z(a, x) \equiv \begin{cases} 1 & |x| \leqslant a/2 \\ 0 & \text{otherwise } a \geqslant 0 \end{cases}$$

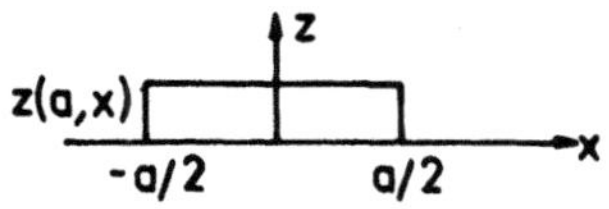

Fig. 16.

h_I and h_E are the densities of the intra- and intervesicular gaps respectively.

The constant building block ρ_M makes this complicated description advantageous: ρ_M can be described by the half period of a Fourier cosine series whose coefficients are the structural parameters. This is analogous to (2.17) and to crystallographic descriptions. The rather lengthy calculation of Q_{ST} according to (3.7), assuming a Gaussian distribution of the intra- and intervesicular intermembrane distances is given in 5.4. The result is:

$$
\begin{aligned}
Q_{ST}(x \geqslant 0) = {} & \\
& \sum_{j=0} \langle n-j\rangle \cdot 2\, \overset{2}{\tilde{\rho}}_M * (H1 * H2)^j \\
& + \sum_{j=0} \langle n-j\rangle \cdot (\rho_{\overline{M}} * \rho_{\overline{M}} * H1) * (H1 * H2)^j \\
& + \sum_{j=0} \langle n-j-1\rangle \cdot (\rho_M * \rho_M * H2) * (H1 * H2)^j \\
& + \sum_{j=0} \langle n-j\rangle \cdot 2\, (\rho_{\overline{M}} * zpI) * (H1 * H2)^j \\
& + \sum_{j=0} \langle n-j-1\rangle \cdot 2\, (\rho_M * zpI * H2) * (H1 * H2)^j \\
& + \sum_{j=0} \langle n-j-1\rangle \cdot 2 \cdot (\rho_M * zpE) * (H1 * H2)^j \qquad (3.10) \\
& + \sum_{j=0} \langle n-j-1\rangle \cdot 2\, (\rho_{\overline{M}} * zpE * H1) * (H1 * H2)^j \\
& + \sum_{j=0} \langle n-j-2\rangle \cdot (zpE * zpE * H1) * (H1 * H2)^j \\
& + \sum_{j=0} \langle n-j-1\rangle \cdot (zpI * zpI * H2) * (H1 * H2)^j \\
& + \sum_{j=0} \langle n-j-1\rangle \cdot 2 \cdot (zpI * zpE) * (H1 * H2)^j \\
& + \langle n\rangle \cdot zqI + \langle n-1\rangle\, zqE
\end{aligned}
$$

The following definitions were introduced:

$$\Phi(x) \equiv \frac{1}{\sqrt{2\pi}} \int_{-\infty}^{x} e^{-\frac{z^2}{2}} dz \qquad \textit{Gaussian error function}$$

$$H1(x) \equiv \langle \delta(x - (l_{2i} - l_{2i-1})) \rangle_{i,s} = \frac{1}{\sqrt{2\pi}\,\sigma_1} e^{-\frac{(x-d)^2}{2\sigma_1^2}}$$

$$H2(x) \equiv \langle \delta(x - (l_{2i+1} - l_{2i})) \rangle_{i,s} = \frac{1}{\sqrt{2\pi}\,\sigma_2} e^{-\frac{(x-d)^2}{2\sigma_2^2}}$$

σ_1, σ_2 are the std. dev. of H1 and H2 respectively

$$zpI \equiv h_I \left\langle \sigma_i^I \cdot Z(|l_{2i} - l_{2i-1} - d|) * \delta\left(x - \frac{l_{2i} - l_{2i-1}}{2}\right)\right\rangle_{i,s}$$

$$= \begin{cases} -h_I\, \Phi\left(\dfrac{x - d/2}{\sigma_1}\right) & \text{for } x < d/2 \\ h_I \left(1 - \Phi\left(\dfrac{x - d/2}{\sigma_1}\right)\right) & \text{for } x > d/2 \end{cases}$$

$$zpE \equiv h_E \left\langle \sigma_i^E \cdot Z(|l_{2i+1} - l_{2i} - d|) * \delta\left(x - \frac{l_{2i+1} - l_{2i}}{2}\right)\right\rangle_{i,s}$$

$$= \begin{cases} -h_E\, \Phi\left(\dfrac{x - d/2}{\sigma_2}\right) & \text{for } x < d/2 \\ h_E \left(1 - \Phi\left(\dfrac{x - d/2}{\sigma_2}\right)\right) & \text{for } x > d/2 \end{cases}$$

$$zqI \equiv h_I^2 \langle \overbrace{Z(|l_{2i} - l_{2i-1} - d|)}^{2} \rangle_{i,s}$$

$$= \frac{2 h_I^2 \sigma_1}{\sqrt{2\pi}} e^{-\frac{x^2}{2\sigma_1^2}} - 2 h_I^2 \cdot x \cdot \Phi\left(\frac{-x}{\sigma_1}\right)$$

$$zqE \equiv h_E^2 \langle \overbrace{Z(|l_{2i+1} - l_{2i} - d|)}^{2} \rangle_{i,s}$$

$$= \frac{2 h_E^2 \sigma_2}{\sqrt{2\pi}} e^{-\frac{x^2}{2\sigma_2^2}} - 2 h_E^2 \cdot x \cdot \Phi\left(\frac{-x}{\sigma_2}\right)$$

Here and in the following chapter $\langle n - j \rangle$ stands for $\langle (n - |j|)_p \rangle$ as defined in Chap. 2.

In this model of the Q_{ST} function, from (3.2) we describe the distance order within the stacks by the two r.v. l_1, l_2 defining the two intermembrane distances. See Fig. 18 for this *membrane concept*.

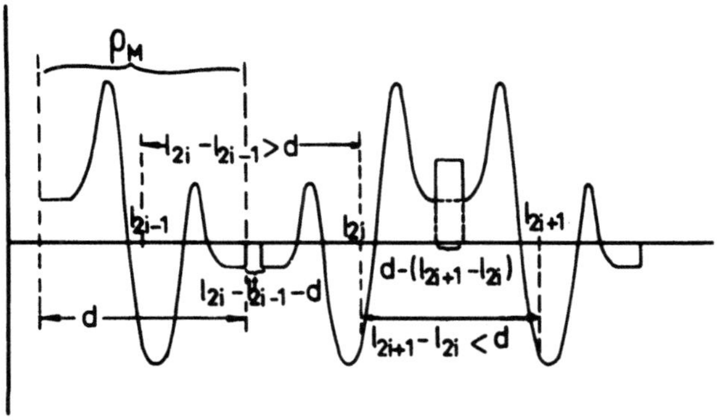

Fig. 17. Layer profile of a stack after substitution of ρ_M showing the location of the artificial gaps and rectangles

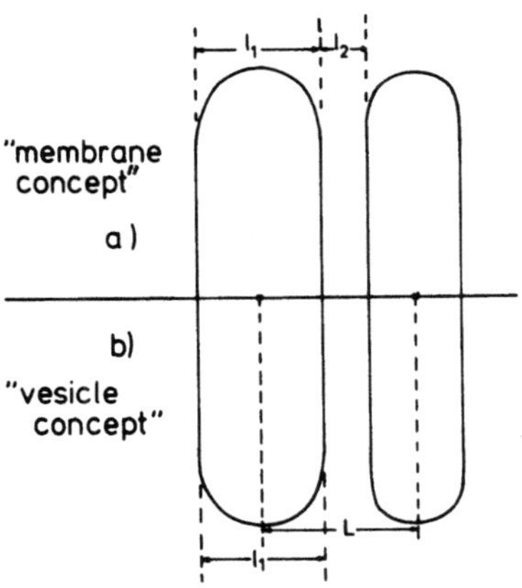

Fig. 18. Comparison of the two possible sets of r.v. characterizing the "membrane concept" and the "vesicle concept" respectively

An alternative concept is to emphasize the whole vesicle as a building block of membrane stacks[34–37, 39]: the distance order within the stacks is described by the following two r.v.: $\mathbf{l}_1$ defined as above and $\mathbf{L}$ the distance between centres of gravity of adjacent vesicles. See Fig. 18 for this *vesicle concept.* In our opinion the membrane concept is a better approach to the true interactions between membranes: interaction forces, which are the physical source for the different distances between membranes, act between membrane surfaces and not between centres of gravity of vesicles. The two concepts differ in the pattern of distance correlation between membranes. An example is the following std. dev. of distance statistics for the membrane concept: $\sigma_1 = 10$ Å; $\sigma_2 = 5$ Å i.e. there is a poor correlation between membranes within a vesicle and a good correlation between membranes of adjacent vesicles. Figure 19 shows the resulting stand.dev. σ_B for distances of two remote membranes:

Fig. 19. $\sigma_A = 5$ Å, $\sigma_B = \sqrt{(10\ \text{Å})^2 + (5\ \text{Å})^2 + (10\ \text{Å})^2} = 15$ Å

The vesicle concept would be incapable of giving such a correlation pattern, since it always gives the same std. dev. for the distances A and B. This seems to us an argument against the vesicle concept, though some biomembrane stacks might obey it.

3.2 Reduction of Q_{ST} to a Simple and Customary Formula for the Special Case of Vanishing Electron Densities of Interspaces

The expression for the Q_{ST}-function (3.10) can be reduced to its first three sums by assuming that for each of the two types of gaps either of the following two statements holds:

1) The electron density of the gap is equal to the mean electron density of the whole stack, i.e. it is zero.
2) The std. dev. of the distance statistics governing the width of the gaps, is negligibly small and can be approximated by a delta function.

In both cases all terms of the Q_{ST} formula containing *zqI, zqE, zpI, zpE* disappear [s. (3.10)], leaving only the first three rows of Q_{ST}. In previous papers we and other authors used this simplified formula to determine the layer profile of membranes[38,41,43,48]. In our evaluations[38,43,48] ρ_M is represented by a sum of Gaussians, whose parameters are varied, until the first three rows of (3.10) fit the experimental Q_{ST}-function. Having measured an intensity I_{exp}, the scattering close to the primary beam shows the background-like Fourier transform of the functions *zqI* and *zqE*. If they are zero, the simplified version of the Q_{ST}-function is applicable to fit the experimental values.

3.3 Problems with the Paracrystalline Plane-Layer Model

It was often difficult to fit experimental Q_{ST}-functions with (3.10). In order to improve the fit, we either had to admit unreasonably large σ_1, σ_2 values (Fig. 20) that are impossible, since membranes cannot penetrate each other, or we had to admit unreasonably short stacks in contradiction with electron microscope pictures.

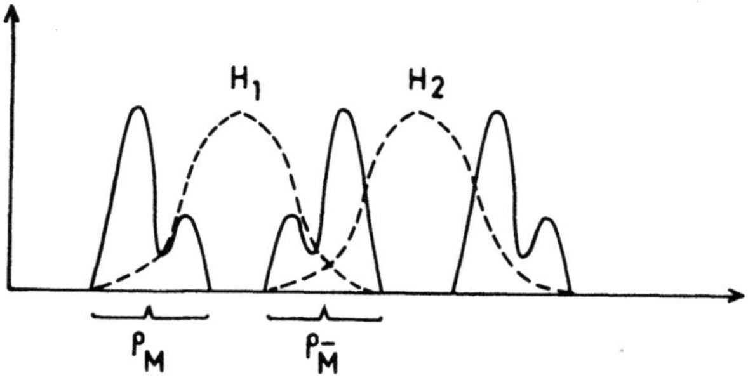

Fig. 20. Part of a membrane stack showing distance statistics H1, H2 whose std. dev. are unreasonably large, so that membranes penetrate each other

From these observations we conclude that the experimental Q_{ST}-functions converge faster on zero with increasing distance than our model Q_{ST}-function. In order to explain this, we investigated the effect of gentle membrane surface undulations with small *"amplitudes"* (≈ 10–100 Å) and large *"wavelengths"* (≈ 500–1000 Å), as occur in most electron microscope pictures of biological membranes. See Fig. 6a for a schematic sketch of such surface undulations. Chap. 4 reports our results.

4 General Paracrystalline Layer Lattice

4.1 Assumptions of the Model

The model is based on the following assumptions:

(4.1) The membrane stacks consist of a large number of vesicles (more than 10)

(4.2) Distances between adjacent membranes of adjacent vesicles (*inter-vesicular distances*) show a probability distribution $H2(l_2)$.
Distances between membranes within a vesicle (*intra-vesicular distances*) show a disorder characterized by a probability distribution $H1(l_1)$, which is uncorrelated with the H2 distribution.

(4.3) The membranes show a wavy surface with small "amplitudes" compared with their "wavelengths" (s. Fig. 6a).

(4.4) The membrane layer profiles are all identical, but are arranged in a sequence of image and mirror image.

4.2 Effect of Membrane Undulations on I_{exp}

In the case of plane membranes, the general equation relating I_{ST} to I_{exp}: (1.12) could be transformed to a simple multiplication of I_{ST} by a factor (s. Chap. 2). We will now demonstrate the failure of this simplification when reasonable undulations in the membrane surface emerge. We begin with the previous case of a plane membrane stack as assumed in Chaps. 2 and 3:

$$Q_{ST} \propto 1\,(R) \qquad I_{ST} \propto \frac{\delta\,(b_R)}{\pi\, b_R} \tag{4.5}$$

Here we have approximated in the Q_{ST}-function the necessarily finite extension of the membrane by an infinite extension and correspondingly in intensity we have concentrated the whole intensity within a small but necessarily finite cylindrical region around the z-axis into a delta-cylinder. This is legitimate, since the reciprocal length of the lateral membrane extension is very small compared with all other details in reciprocal space. In this plane membrane model, the general integral (1.12) is equivalent (up to a constant) to multiplication of I_{ST} with $1/b_z^2$, as was derived in Chap. 2 (s. Fig. 21).

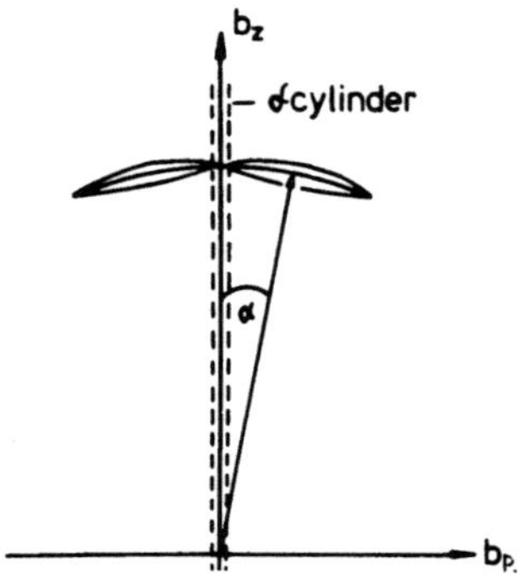

Fig. 21. Schematic of integral (1.12) in the case of plane membranes. The arrow indicates the patch of integration, its width the function $h(\alpha) \cdot \sin(\alpha)$

Fig. 22. Formation of undulations in a plane membrane

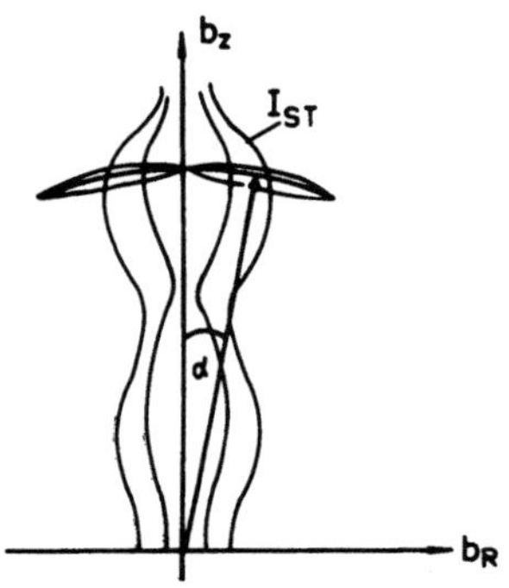

Fig. 23. Schematic of integral (1.12) in the case of undulated membranes

We now assume, that undulations of the membranes emerge, as shown schematically in Fig. 22.

As a consequence of these undulations, the membrane electron density as well as the Q_{ST}-function will change its approximate radial dependence $\simeq 1$ (R) and begin to fade off in radial direction, as can be seen from Fig. 22. Consequently the intensity I_{ST} will float across the initial δ-cylinder, as is shown schematically in Fig. 23.

As long as the radial extension of I_{ST} is much smaller than $\sigma_\alpha \cdot b_z$ [σ_α is the std. dev. of $h(\alpha)$] the Lorentz factor $1/b_z^2$ will be a reasonable approximation for (1.12), but in the general case it is invalid and we have to know the exact three-dimensional form of I_{ST} in reciprocal space in order to evaluate I_{exp}. Before we evaluate the general three-dimensional expression of I_{ST} via Q_{ST}, we will demonstrate an additional effect of membrane undulations on the integral (1.12). For the sake of simplicity we assume "stacks" consisting of only one vesicle with a delta-like distribution of inter-membrane distances. In Chaps. 2 and 3 I_{ST} was restricted to a delta-cylinder around the b_z-axis and as a consequence, all our knowledge of membrane structure was based on the projection of the vesicle-layer density onto the z-axis:

$$I_{ST}(0, 0, b_z) = F\langle \widetilde{P_z \rho_0}^2 \rangle_s \quad (4.6)$$

This projection $P_z\rho_0$ should show serious effects from emerging undulations on the membrane surface, see Fig. 24.

In order to investigate this phenomenon quantitatively, we now give a mathematical model to calculate the projection $P_z\rho_0$ of a wavy vesicle on its stacking axis. The layer vesicle can be thought of as divided into infinitesimally narrow pillars parallel to the stacking axis and having quadratic cross sections as shown in Fig. 25.

If ρ_0 is the layer profile of the vesicle, the z-projection can be written:

$$[P_z\rho_0](z) = \int dz' \int d\alpha \rho_0((z - z') \cos\alpha)\, \eta(z', \alpha) \quad (4.7)$$

Where $\eta(z', \alpha)$ is the frequency of occurrence of pillars with the parameters (z', α). For a cap of a sphere with height h and curvature 1/R (s. Fig. 26) $\eta_R^h(z', \alpha)$ can be derived as shown in 5.5.

As a rough model for an undulated surface, we use a distribution of spherical caps with Gaussian distribution of the parameters h and R. Such an undulated model-membrane could look like Fig. 27 when viewed from above.

Disregarding the hatched areas of the surface, the density function $\eta_{\langle R \rangle}^{\langle h \rangle}(z', \alpha)$ of an ensemble of spherical caps is simply a superposition of densities $\eta_R^h(z', \alpha)$ of spherical caps, as defined above. This calculation is given in 5.5. Figure 28 shows model calculations of $P_z\rho_0$ for increasing "amplitudes" $\langle \mathbf{h} \rangle$ of the caps evaluated

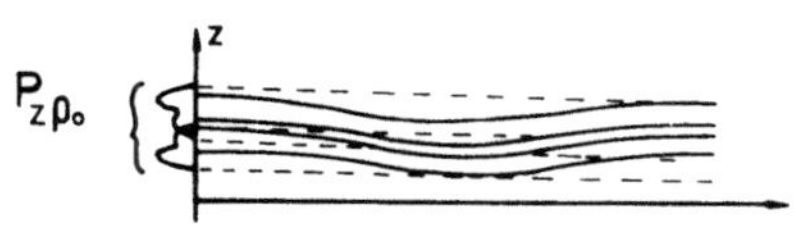

Fig. 24. Effect of undulations on the projection of the layer-density on the stacking axis

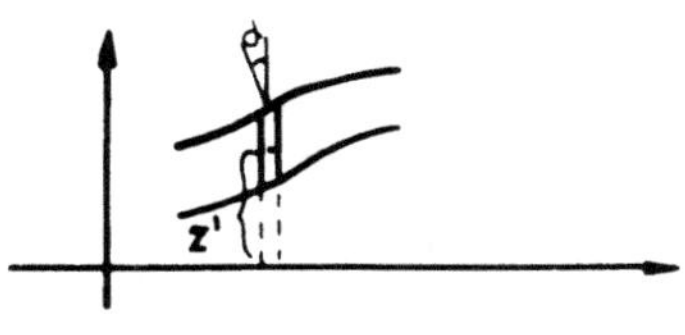

Fig. 25.

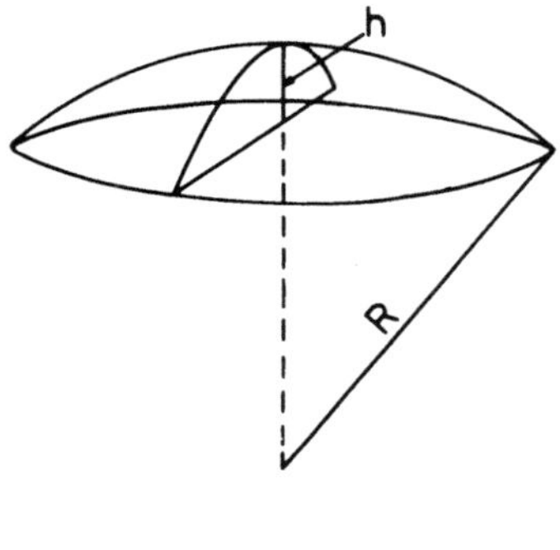

Fig. 26. Cap of a sphere characterized by height h and curvature 1/R

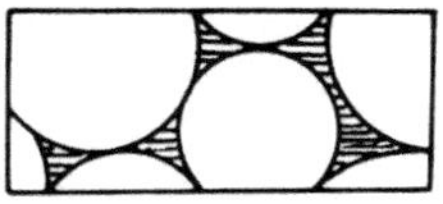

Fig. 27. Section of an undulated membrane surface composed of spherical caps when viewed from above

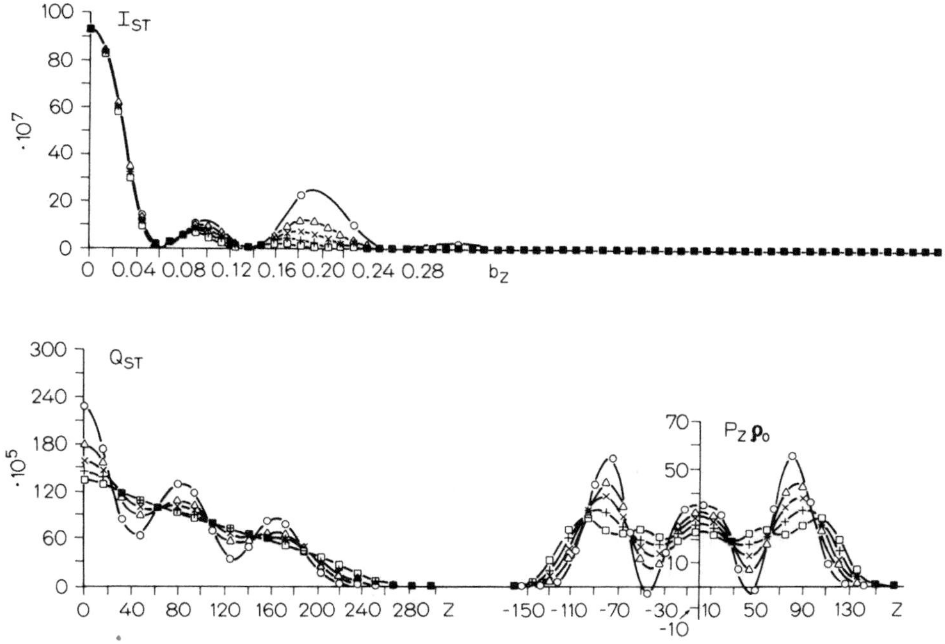

Fig. 28. Undulations of a single vesicle: Effects on the projection of the layer-density, the projection of the Q_{ST}-function and the intensity $I_{ST}(0, 0, b_z)$, when the mean amplitude $\langle h \rangle$ of the undulations is increased: ○ : $\langle h \rangle = 2$ Å; ▲ : $\langle h \rangle = 10$ Å; x : $\langle h \rangle = 20$ Å; + : $\langle h \rangle = 30$ Å; □ : $\langle h \rangle = 40$ Å

with (4.7) and the function $\eta_{\langle R \rangle}^{\langle h \rangle}(z', \alpha)$. The Q_{ST}-function and the intensity $I_{ST}(0, 0, b_z)$ are also calculated.

As expected, the projection loses all characteristics of the profile when the mean amplitude $\langle \mathbf{h} \rangle$ of the surface undulations reaches the order of magnitude of the wavelength Δ of the most prominent Fourier-waves of ρ_0. Typical Δ's for membranes are about 20 Å. Electron microscope pictures indicate surface "amplitudes" equal to or larger than 20 Å. In practical cases therefore $I_{ST}(0, 0, b_z)$ will be greatly affected by the membrane surface undulations and the contribution of the δ-cylinder to the integral (1.12), i.e. to the measured intensity I_{exp}, is considerably reduced. However, beyond the delta cylinder we fortunately obtain new contributions to I_{exp} giving measurable I_{exp} values even for "amplitudes" of surface undulations greater than Δ. Consequently the projection $P_z\rho_M$ will now deliver its previous role as ultimate information of membrane structure to the membrane layer profile ρ_M.

Up to now we have examined the effect of undulations mainly by examining the Q_{ST}-function in real space and the effect of the flower-bunch statistic by Eq. (1.12) in reciprocal space. Discussion of both effects is easier, when they are judged from the view of real space.

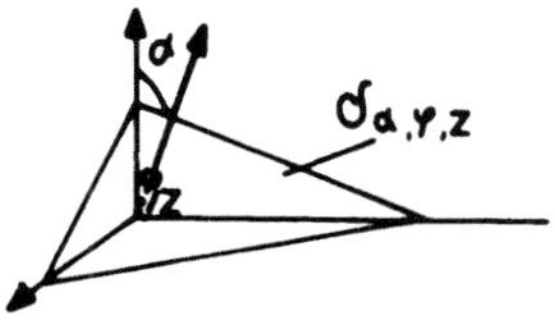

Fig. 29.

We will therefore now give a vivid description of how $\mathbf{Q}_{ST}$ is related to $Q_{exp} \equiv F^{-1}\{I_{exp}\}$. Evidently we have to take the Fourier transform of (1.12):

$$Q_{exp}(z) = \int I_{exp}(b_z)\, e^{2\pi i b_z \cdot z}\, db_z$$

$$= \int_0^{2\pi} d\varphi \int_0^{\pi} h(\alpha) \sin\alpha \int I_{ST}(b_z, \alpha)\, e^{2\pi i b_z \cdot z}\, db_z\, d\alpha \tag{4.8}$$

using the projection theorem (s. 1.3):

$$= \int_0^{2\pi} d\varphi \int_0^{\pi} h(\alpha) \sin\alpha\, \{P_{\alpha,\varphi}\, \mathbf{Q}_{ST}(\vec{x})\}(z)\, d\alpha$$

$P_{\alpha,\varphi}\, \mathbf{Q}_{ST}$ is the projection of $\mathbf{Q}_{ST}$ onto an axis with spatial orientation α, φ (φ arbitrary) relative to the stacking axis $\hat{z}$. We can express it by a volume integral containing a delta function:

$$Q_{exp}(z) = \int_0^{\pi} d\alpha\, h(\alpha) \sin\alpha\, \{\int d\varphi \int d\tau\, \delta_{\alpha,\varphi,z}(\vec{x})\, \mathbf{Q}_{ST}(\vec{x})\} \tag{4.9}$$

where $\delta_{\alpha,\varphi,z}$ is a delta-plane, whose normal is given by α, φ and whose nearest distance from the origin is z, as shown in Fig. 29.

Commuting integrals in (4.9) we get:

$$Q_{exp}(z) = \int d\tau\, \{\int_0^{2\pi} d\varphi \int_0^{\pi} h(\alpha) \sin\alpha\, d\alpha\, \delta_{\alpha,\varphi,z}(\vec{x})\, \mathbf{Q}_{ST}(\vec{x})$$

$$\equiv \int d\tau\, g_z(r, \theta)\, \mathbf{Q}_{ST}(r, \theta) \tag{4.10}$$

with spherical coordinates r, θ.

Visually g_z is a superposition of delta-planes, which all touch a sphere of radius z and whose normals are distributed according to $h(\alpha)$.

Solution of the integral

$$g_z(r, \theta) = \int d\varphi \int d\alpha\, h(\alpha) \sin\alpha\, \delta_{\alpha,\varphi,z}(\vec{x})$$

is possible by a method by Deas[51a)]: we have to calculate the radial expansion functions of the spherical harmonics for a delta-plane with distance z from the origin and extending perpendicular to the z-axis:

$$G_n^m(r) = \frac{2n+1}{4\pi}\frac{(n-m)!}{(n+m)!} \int_0^{\pi} d\theta \int_0^{2\pi} d\varphi \cdot \delta(z - r\cos\theta)\, P_n^m(\cos\theta)\, e^{-im\varphi} \sin\theta$$

we find:

$$G_n^0(r) = \begin{cases} \frac{2n+1}{2} P_n\left(\frac{z}{r}\right) & \text{for } r \geqslant z; \\ 0 & \text{otherwise} \end{cases} \qquad G_n^{m\neq 0} = 0$$

Inserting this result into the end formula of Deas:

$$g_z(r,\theta) = \begin{cases} \sum_{\nu=0} h_\nu \cdot \frac{1}{r} P_\nu\left(\frac{z}{r}\right) P_\nu(\cos\theta) & \text{for } r \geqslant z \\ 0 & \text{otherwise} \end{cases}$$

$$h_\nu \equiv \frac{2\nu+1}{2} \int_0^{\pi} h(\alpha)\, P_\nu(\cos\alpha) \sin\alpha \, d\alpha$$

Figure 30 shows g_z for a Gaussian $h(\alpha)$ of typical dimension and for z-values $z = 0$ Å, 75 Å, 150 Å.

The functions g_z are neither continuous nor integrable, since they are formed by a superposition of delta-planes. Nevertheless the integral (4.10) exists, since Q_{ST} converges to zero in all directions (even when membranes are plane). As we conclude from Fig. 30, Q_{exp} is formed by integrating the product of Q_{ST} with a special weight-function g_z, which emphasizes the regions of Q_{ST} near the z-axis.

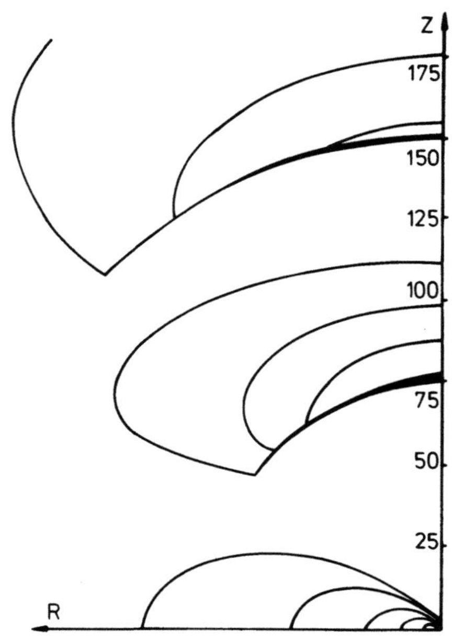

Fig. 30. The function g_z for parameter values $z = 0$ Å, 75 Å, 150 Å and the st d. dev. of $h(a)$; $\sigma_a = 0.5$ (radians)

The integral (4.10) becomes simple again, when membranes have a plane surface, i.e. when Q_{ST} is of the form assumed in Chaps. 2 and 3: (4.5). In this case $\delta_{\alpha,\varphi,z}$ contributes to Q_{exp} the integral:

$$\int \delta_{\alpha,\varphi,z}(\vec{x})\, Q_{ST}(\vec{x})\, d\tau$$

Except for the case $\alpha = 0$ all these contributions disappear as we can conclude from the projection theorem:

$$F\{\int \delta_{\alpha,\varphi,z}(\vec{x})\, Q_{ST}(\vec{x})\, d\tau\} = I_{ST}(b_r, \alpha) \simeq \frac{\delta(b_r \sin\alpha)}{\pi b_r \sin\alpha}\, I_{ST}(0, 0, b_r)$$

We conclude from the visualization of the integral (4.10), that the correct evaluation of Q_{exp} or I_{exp} must take into account the shape of the Q_{ST}-function in the neighbourhood of the z-axis, which is determined by the wavy membrane structure.

4.3 Evaluation of the Q_{ST}-Function for the General Paracrystalline Layer Lattice

According to (4.3) we assume that the membranes undulate gently above and below a midplane, whose normal is parallel to the stacking axis and which intersects the z-axis at a distance l_i from $z = 0$. If $s_i(R, \varphi)$ gives the height of the membrane relative to its midplane, then $\delta(z - (l_i + s_i(R, \varphi)))$ is a delta-plane, nestling at the centre of our wavy membrane. In the following we will use ρ_M not just for the layer profile $\rho_M(z)$, but rather for an infinitesimal pillar out of it:

$$\rho_M \equiv \rho_M(z) * \frac{\delta(R)}{\pi R}$$

Considering the small angles of the surface normals with the stacking axis which follow from (4.3), it should be a good approximation for a stack of wavy membranes to write:

$$\rho_s = \sum_{i=1}^{n(s)} \{\rho_M * \delta(z - (l_{2i-1} + s_{2i-1})) + \rho_{\bar{M}} * \delta(z - (l_{2i} + s_{2i}))\}$$

$$+ \left\{ h_I\, \sigma_i^I \cdot Z\left(|l_{2i} + s_{2i} - l_{2i-1} - s_{2i-1} - d|, z - \frac{l_{2i} + s_{2i} + l_{2i-1} + s_{2i-1}}{2} \right) \right\} \quad (4.11)$$

$$+ \left\{ h_E\, \sigma_i^E \cdot Z\left(|l_{2i+1} + s_{2i+1} - l_{2i} - s_{2i} - d|, z - \frac{l_{2i} + s_{2i} + l_{2i+1} + s_{2i+1}}{2} \right) \right\}$$

The arguments (R, φ) at s and the subscript s at σ_i^I, σ_i^E, s_i, l_i were omitted for simplicity.

$\sigma_i^E \equiv \text{sgn}\,(l_{2i} + s_{2i} - l_{2i-1} - s_{2i-1} - d)$

$\sigma_i^I \equiv \text{sgn}\,(l_{2i+1} + s_{2i+1} - l_{2i} - s_{2i} - d)$

In the case of a plane membrane stack $\langle \overset{2}{\tilde{\rho}}_s \rangle_s$ was a function only of z, now with membrane undulations it will depend upon z and R (cylindrical coordinates) and consequently have a cylindrical symmetry. Let us first calculate $\overset{2}{\tilde{\rho}}_s$ on the z-axis (R = 0):

$$\overset{2}{\tilde{\rho}}_s\,(R=0, z) = \int d\bar{\varphi} \int d\bar{R} \cdot \bar{R} \int d\bar{z}\, \rho_s\,(\bar{\varphi}, \bar{R}, \bar{z})\, \rho_s\,(\bar{\varphi}, \bar{R}, \bar{z} - z)$$

Inserting (4.11) and substituting $\bar{l}_i = l_i + s_i$, we see, that the inner integral

$$\int d\bar{z}\, \rho_s\,(\bar{\varphi}, \bar{R}, \bar{z})\, \rho_s\,(\bar{\varphi}, \bar{R}, \bar{z} - z)$$

is of the same type as $\overset{2}{\tilde{\rho}}_s$ in the case of plane membranes, as was given in Chap. 3. The two external integrals $\int d\bar{\varphi} \int d\bar{R} \cdot \bar{R}$ have the same effect, as taking a mean of convolution squares over a sample of plane membrane stacks. Finally we have to evaluate the mean for all stacks $\langle \overset{2}{\tilde{\rho}}_s\,(R = 0, z)\rangle_s$, which gives an additional summation of the same general kind as above. We conclude that $\mathbf{Q}_{ST}\,(0, 0, z) = \langle \overset{2}{\tilde{\rho}}_s\,(0, 0, z)\rangle_s$ is of the same general form (3.10), but here the st d.dev. σ_1 and σ_2 not only express the $l_i - l_{i-1}$ fluctuations but also these arising from the undulations s_i.

In order to calculate Q_{exp} from $\mathbf{Q}_{ST}$, we have to know the latter function in the neighbourhood of the z-axis, as was shown earlier. We will have to investigate the single convolution products of (3.10) for their R dependence, i.e. for removal from the z-axis. Alternatively it would be possible, but laborious and not very instructive, to calculate $\overset{2}{\tilde{\rho}}_s(R, z)$ using the general formula (4.11). In most cases Q_{ST} converges to zero after having reached twice the Bragg period. It will therefore be sufficient to investigate the convolution products of only the first Bragg period. Having done this, the general case will be evident.

The convolution products of the first Bragg period follow from (3.10) for j = 0:

$$\begin{aligned}
&2\,\langle n\,(s)\rangle\, \overset{2}{\tilde{\rho}}_M + \langle n\rangle\, zqI + \langle n-1\rangle\, zqE \\
&+ \langle n\rangle \cdot \rho_{\bar{M}} * \rho_{\bar{M}} * H1 + \langle n-1\rangle \cdot \rho_M * \rho_M * H2 + 2\,\langle n\rangle \cdot \rho_{\bar{M}} * zpI \\
&+ 2\,\langle n-1\rangle \cdot \rho_M * zpI * H2 + 2\,\langle n-1\rangle \cdot \rho_M * zpE \\
&+ 2\,\langle n-1\rangle \cdot \rho_{\bar{M}} * zpE * H1 + \langle n-2\rangle \cdot zpE * zpE * H1 \\
&+ \langle n-1\rangle \cdot zpI * zpI * H2 + 2\,\langle n-1\rangle \cdot zpI * zpE
\end{aligned} \tag{4.12}$$

These convolution products can be classed as 3 different R-dependences:

1) Consider $\overset{2}{\tilde{\rho}}_M$ as a typical example. Evidently, the (R, z) dependence of the convolution square of a wavy building block ρ_M will be:

$$\overset{2}{\tilde{\rho}}_M * \langle \delta\,(z - s_{2i}\,(R,\varphi)) * \delta\,(z + s_{\overline{2i}}\,(R,\varphi)) + \delta\,(z - s_{2i-1}\,(R,\varphi)) * \delta\,(z + s_{\overline{2i}-1}\,(R,\varphi))\rangle_{i,s}$$

Where we have omitted the subscript s from *s* for simplicity. Mathematically the convolution products of δ-surfaces can be evaluated as described by Papoulis[49d]. We simply assume that the mean value above is of the following general form[42]:

$$\frac{1}{\sqrt{2\pi}\,\sigma(R)}\; e^{-\frac{z^2}{2\sigma^2(R)}}$$

where δ (R) is an increasing function starting from zero for R = 0

This is a function of cylindrical symmetry that starts with a delta function for R = 0 and for increasing R smears into a Gaussian with increasing st d.dev., in this way describing the *lateral correlation* of the membrane surface with itself.

Qualitatively we expect σ (R) as drawn schematically in Fig. 31.

At large R-values, σ (R) reaches the order of magnitude of the surface wave amplitudes. From electron microscopy we know that in this case $\sigma(R) \geqslant \Delta$($\Delta$ is the wavelength of the most prominent Fourier wave in the Fourier series of ρ_M) and we conclude:

$$\overset{2}{\tilde{\rho}}_M * \frac{1}{\sqrt{2\pi}\,\sigma\,(R \to \infty)}\; e^{-\frac{z^2}{2\sigma^2(R\to\infty)}} \approx 0$$

We therefore expect, that the region of R, having σ (R) $\leqslant \Delta$ and nondisappearing convolution products, consists only of the inner part of Fig. 31 and we thus assume a linear model:

$$\sigma(R) \cong A \cdot R$$

Definition:

$$H_i^w\,(R, z) \equiv \frac{1}{\sqrt{2\pi}\,A_i^w \cdot R}\; e^{-\frac{z^2}{2(A_i^w \cdot R)^2}} \qquad (4.13)$$

The same lateral correlation pattern as for $\overset{2}{\tilde{\rho}}_M$ is also found for the convolution products $\rho_{\overline{M}} * zpE$, $\rho_M * zpE$, $zpI * zpE$ since the inner and outer gaps are firmly attached to the membranes. We also assume that the radial dependence of *ZQI* and *ZQE* can be approximated by H_I^w.

2) We now investigate a convolution product involving two different membrane surfaces: $\rho_{\overline{M}} * \rho_{\overline{M}}$. Here the two neighbouring membranes of one vesicle are convoluted.

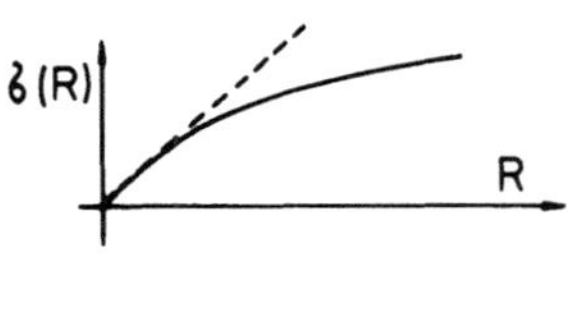

Fig. 31. std. dev. σ (R) of lateral correlation of membrane surfaces as a function of distance R and its linear approximation

Evidently their surfaces show a different lateral correlation, than each surface with itself. Their lateral correlation will be:

$$\langle\!\langle \delta\,(z-(l_{2i}+s_{2i})) * \delta\,(z+(l_{2i-1}+s_{\overline{2i}-1}))\rangle\!\rangle_{i,s} \tag{4.14}$$

Three possible cases must be discussed:

a) The surfaces of membranes are totally uncorrelated, see Fig. 32:

$$\int (s_i\,(R,\varphi)-s_{i-1}\,(R,\varphi))^2 \cdot R dR d\varphi \approx 2 \int s_i^2(R,\varphi) \cdot R dR d\varphi$$

consequently $\sigma\,(R) = \text{const.} > \Delta$ for all R. Keeping in mind electron microscopic pictures, convolution products of this case should disappear effectively everywhere.

b) The membrane surfaces show total correlation, see Fig. 33:

$$\int (s_i\,(R,\varphi)-s_{i-1}\,(R,\varphi))^2 \cdot R dR d\varphi = 0$$

or $s_i = s_{i-1}$. We can transform (4.14) into:

$$\langle \delta\,(z-s_{2i}) * \delta(z+s_{\overline{2i}-1})\rangle_{i,s} * \left\langle \delta\,(z-(l_{2i}-l_{2i-1}))\cdot\frac{\delta\,(R)}{\pi R}\right\rangle_{i,s}$$

$$= H_1^w\,(R,z) * \left[\frac{1}{\sqrt{2\pi}\,\sigma_1}\,e^{-\frac{(z-d)^2}{2\sigma_1^2}} \cdot \frac{\delta\,(R)}{\pi R}\right] = H_1^w\,(R,z) * H1 \cdot \frac{\delta\,(R)}{\pi R}$$

c) The membranes show partial correlation, as shown schematically in Fig. 34.

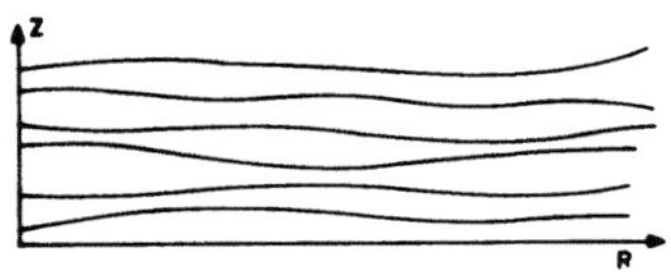

Fig. 32. Schematic representation of a stack with uncorrelated membrane surfaces

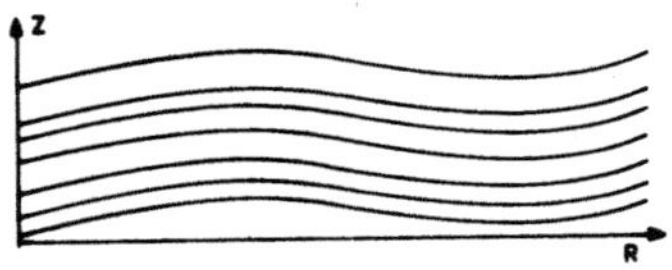

Fig. 33. Schematic representation of a stack with totally correlated membrane surfaces

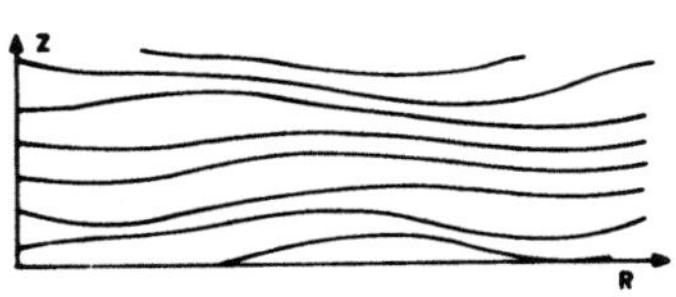

Fig. 34. Schematic representation of a stack with partially correlated membrane surfaces

In this case we assume that the following approximation is justified:

$$\langle \delta (z - (l_{2i} + s_{2i})) * \delta (z + (l_{2i-1} + s_{\overline{2i}-1}))\rangle_{i,s}$$

$$= \frac{1}{\sqrt{2\pi}\, A_2^w \cdot R}\, e^{-\frac{z^2}{2(A_2^w \cdot R)^2}} * \left[\frac{1}{\sqrt{2\pi}\,\sigma_1}\, e^{-\frac{(z-d)^2}{2\sigma_1^2}} \cdot \frac{\delta (R)}{\pi R}\right] \tag{4.15}$$

$$= H_2^w * \left[H1 \cdot \frac{\delta (R)}{\pi R}\right]$$

As opposed to case b), σ_1 now contains not only the st d.dev. of $l_{2i} - l_{2i-1}$, but also a contribution from the disorder caused by partial uncorrelation of the surfaces. Assuming $A_2^w = A_1^w$, we get case b). But in contrast to case a) σ_1 does not contain all the distance disorder caused by the surface undulations. We obtain case a) if we substitute $A_2^w = 0$ and take σ_1 as large as the membrane "amplitudes".

The same lateral correlation patterns as found for $\rho_{\overline{M}} * \rho_{\overline{M}}$ is common to all convolution products, whose multiplicands are separated by the r.v. l_1:

$$\rho_{\overline{M}} * \rho_{\overline{M}} * H1,\ zpE * zpE * H1,\ \rho_{\overline{M}} * zpE * H1$$

3) An other lateral correlation, described by a function H_3^w, can be expected for all convolution products influenced by the r.v. l_2:

$$\rho_M * \rho_M * H2,\ zpI * zpI * H2,\ \rho_M * zpI * H2$$

Generalizing these results to all terms of the first two periods of the Q_{ST}-function (j = 0, j = 1 in (3.10)), we find:

$$Q_{ST}(R, z) = \langle \overset{2}{\tilde{\rho}_s} \rangle_s \tag{4.16a}$$

$$= B_0 (R, z) * H_1^w (R, z)$$

$$+ B_1 (R, z) * H_1^w (R, z) + \cdots + B_6 (R, z) * H_6^w (R, z)$$

$$+ [B_1 (R, z) * H_1^w (R, z) + \cdots + B_6 (R, z) * H_6^w (R, z)]^-$$

$$B_0 (R, z) \equiv \left[\langle n\rangle \cdot ZQI + \langle n-1\rangle\, ZQE - 2\,\langle n\rangle \overset{2}{\tilde{\rho}_M}\right] \cdot \frac{\delta (R)}{\pi R}$$

$$B_1 (R, z) \equiv c_1 (0) \qquad B_4 (R, z) \equiv c_1 (1) * H1 * H2$$

$$B_2 (R, z) \equiv c_2 (0) \qquad B_5 (R, z) \equiv c_2 (1) * H1 * H2$$

$$B_3 (R, z) \equiv c_3 (0) \qquad B_6 (R, z) \equiv c_3 (1) * H1 * H2$$

$$c_1\,(g) \equiv \Big[2\,\langle n-g\rangle \overset{2}{\tilde{\rho}}_M + \langle n-g\rangle \cdot \rho_{\overline{M}} * zpI$$

$$+ 2\,\langle n-g-1\rangle\, \rho_M * zpE + 2\,\langle n-g-1\rangle\, zpI * zpE \Big] \cdot \frac{\delta\,(R)}{\pi \cdot R}$$

$$c_2\,(g) \equiv \Big[\langle n-g\rangle \cdot \rho_{\overline{M}} * \rho_{\overline{M}} * H1 + \langle n-g-2\rangle \cdot zpE * zpE * H1$$

$$+ 2\,\langle n-g-1\rangle \cdot \rho_{\overline{M}} * zpE * H1 \Big] \cdot \frac{\delta\,(R)}{\pi R}$$

$$c_3\,(g) \equiv \Big[\langle n-g-1\rangle \cdot \rho_M * \rho_M * H2 + \langle n-g-1\rangle\, zpI * zpI * H2$$

$$+ 2\,\langle n-g-1\rangle \cdot \rho_M * zpI * H2 \Big] \cdot \frac{\delta\,(R)}{\pi R}$$

Here $\langle n-m\rangle$ stands for $\langle (n-|m|)_p\rangle$ as defined in Chap. 2. I_{exp} follows from $\mathbf{Q}_{ST}$ by:

$$I_{exp}\,(b_z) = \int_0^{\pi} h\,(\alpha) \sin\alpha\, [F\,\mathbf{Q}_{ST}]\,(b_z, \alpha)\, d\alpha \qquad (4.16b)$$

4.4 Influence of Undulated Membrane Surfaces on Q_{exp} and I_{exp}

Q-functions of many biomembranes converge to zero for $z > 2 \cdot \langle L \rangle$ ($\langle L \rangle$ = Bragg period). In contrast, electron microscopy of the same stacks usually shows a much larger number of vesicles. As was mentioned in Chap. 3, the paracrystalline plane-layer lattice model cannot account for this fast convergence of the Q-function to zero alone. The more general formulas (4.16) should be able to overcome this difficulty by their new concepts:

a) of the weightening function g_z

b) of an undulated surface of the membranes

The general paracrystalline layer-lattice model with partial correlations of the surfaces can give the following explanation:

The decreasing lateral correlation of the membrane surfaces gives a Q_{ST}-function as shown schematically in Fig. 35.

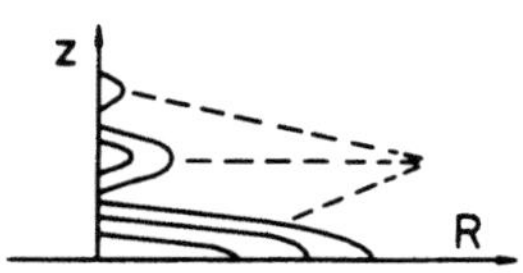

Fig. 35. Rapid radial fading of convolution products of remote membranes by decreasing correlation of membrane surfaces

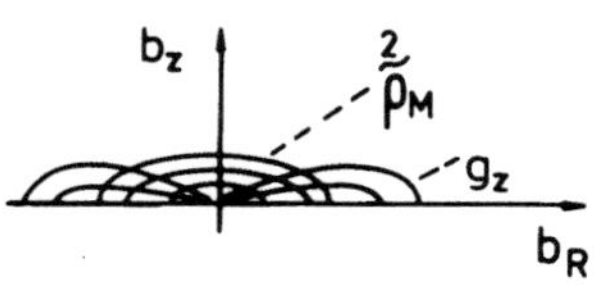

Fig. 36. Schematic superposition of $\overset{2}{\tilde{\rho}}_M$ and $g_z = 0$

The larger the distance z, the smaller will be the integral (4.10) by the radial decrease of Q_{ST} and the spreading of g_z (Fig. 30).

The main preparative task for the membrane stackings to be investigated by small-angle X-ray scattering should therefore be to prepare stackings with partial correlations of surface forms.

With the in vivo water content, most membrane preparations do not show X-ray scattering with distinct reflexes, and at best only a continuous scattering is observed.

We propose the following explanation on the basis of surface correlations: at the in vivo water content, the membranes are very wavy with relatively large intermembrane distances and consequently poor correlation of membrane surfaces. Q_{ST} therefore contains at best the convolution squares $\tilde{\rho}_M^{\,2}$ and zqI and zqE, which rapidly smears off radially. Superposition of Q_{ST} and g_z in this case are drawn schematically in Fig. 36.

I_{exp} therefore only shows a faint, continuous scattering mainly near $b_z = 0$. At descreasing water content, under suitable geometrical conditions (suspended drop, as shown in Fig. 37), the intermembrane distances shrink and the surfaces become less wavy. Owing to the smaller intermembrane spaces, surface correlations must arise. Therefore convolution products of neighbouring membranes emerge and we obtain distinct reflexes, indicating coherence regions of two or more vesicles.

We have tried to set up a model for these effects with the program to be described later, that uses (4.16) to evaluate I_{exp}: Table 1 gives model-parameter values for 4 intensities $I_4, \ldots, I_1$ corresponding to increasing surface correlation. We have disregarded the intermembrane distance shrinkage in order to simplify the diagram.

The I_{exp}-functions following from these parameters for a special membrane layer profile are shown in Fig. 38.

The negative intensity values in I_1 indicate that in this case we have either chosen unreasonably small values for σ_1, σ_2 and A_i^W or, what is an equivalent explana-

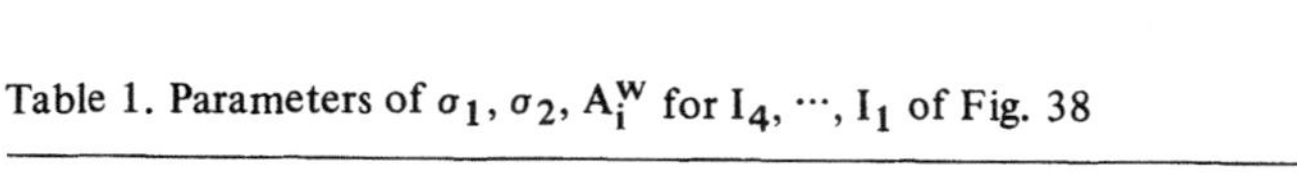

High water content   Low water content

Fig. 37. Suspended drop

Table 1. Parameters of σ_1, σ_2, A_i^W for $I_4, \cdots, I_1$ of Fig. 38

	σ_1	σ_2	A_1^W	A_2^W	A_3^W	A_4^W	A_5^W	A_6^W	A_7^W	A_8^W	A_9^W
I_1	4.	5.	0.1	0.12	0.12	0.22	0.24	0.24	0.34	0.36	0.36
I_2	6.	7.25	0.16	0.19	0.19	0.34	0.38	0.38			
I_3	8.	9.5	0.22	0.32	0.32						
I_4	10.	12.	0.26	0.36	0.36						

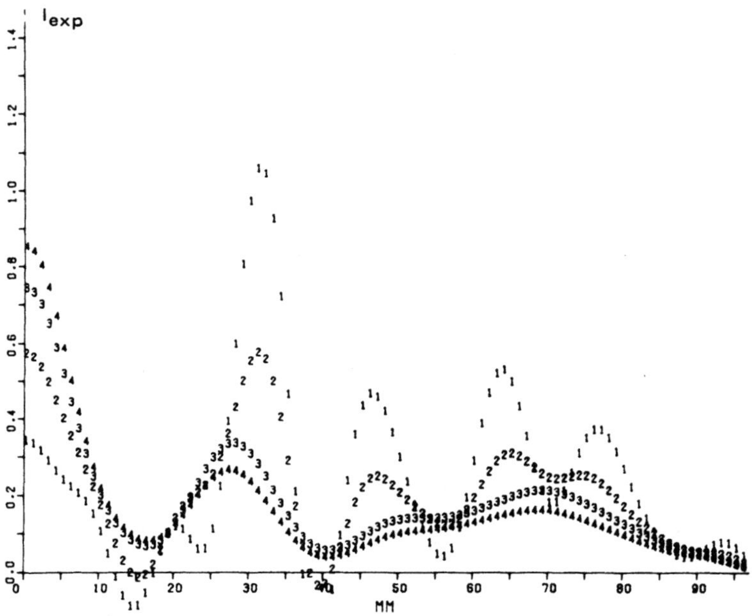

Fig. 38. Model surface correlation effect

tion, that the Q-function did not converge entirely to zero within the three Bragg periods, we have allowed for in the calculation. The Fourier transform of such a truncated Q-function can have negative values.

4.5 A Procedure to Evaluate the Layer Profile Based on the General Paracrystalline Layer Lattice Model

By introducing a suitable set of parameters into (4.16) in order to represent ρ_M, H1, H2, H_i^w, these equations can be transformed into a system on nonlinear equations:

For ρ_M we substitute:

$$\rho_M^{per}(z) \equiv \sum_{k=1}^{NC} b_k \cos\left(2\pi \frac{z}{2d} \cdot k\right)$$

$$\rho_M(z) \equiv \begin{cases} \rho_M^{per}(z - d/2) & -d/2 \leqslant z \leqslant d/2 \\ 0 & \text{otherwise} \end{cases} \tag{4.17}$$

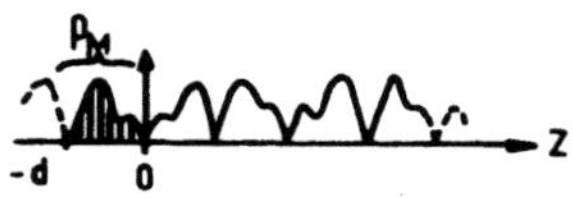

Fig. 39. Schematic drawing of the serieş ρ_M^{per}

For H1, H2 and H_i^w we use the parameters σ_1, σ_2 and A_i^w respectively. For the flower-Bunch statistics we adopt a Gaussian, whose st d.dev. can be measured experimentally with reasonable accuracy. In the same way the Bragg-period $\langle \mathbf{L} \rangle$ and $d = \langle \mathbf{L} \rangle/2$ can be evaluated directly by experiment.

$\langle n - j \rangle$ is estimated from electron microscopic pictures. We therefore define the following set of unknown structural parameters:

$$\{p_i\} \equiv (b_1, \cdots, b_{NC}, \sigma_1, \sigma_2, A_1^w, \cdots) \tag{4.18}$$

Using Shannon's theorem, the intensity $I_{exp}(b_z)$ can be represented by a suitable set of equidistant values I_m of this function.

In this way Eqs. (4.16) are equivalent to a set of nonlinear equations:

$$I_m^{Mod} = f_m(b_1, \cdots, b_{NC}, \sigma_1, \sigma_2, A_1^w, \cdots)\ m = 1, \cdots, N \tag{4.19}$$

The explicit derivation is given in 5.6.

From the experimental scattering intensity we get Shannon's points I_m^{exp}. In order to determine the unknown layer profile of the membranes under investigation, we have to solve Eqs. (4.19) for the unknown structural parameters. Guided by earlier applications, we decided to use Baumeisters program[58)] applying Marquardts least-square fit procedure[53, 54)], which will now be briefly outlined.

For every given set $\{p_i\}$ of structural parameters, a Φ-*value* is obtained:

$$\Phi(\{p_i\}) = \sum_{m=1}^{N} [I_m^{exp} - I_m^{Mod}(\{p_i\})]^2 \tag{4.20}$$

In this way a function Φ is defined, which we describe as the Φ-*surface above the parameter space*. Marquardts procedure starts with a set of inferred initial values: $\{p_i^g\}$ and tends to the point $\{p_i^l\}$ with the nearest attainable local minimum of the Φ-surface. In Fig. 40 we try to give a picture of the Φ-surface, using a two-dimensional parameter space (p_1, p_2).

We know from experience that the Φ-surface exhibits a number of local minima. The minima above the subspace of the Fourier-coefficients b_k (or Shannon points of a kind of artificial structure factor) are of special importance to us. We find a minimum to each of the 2^{NC} phase sets, as can be expected by analogy with classical crystallography. Only 2^{NC-1} of these phases are essentially different, since changing all phases is equivalent to substitution of $-\rho_M$ for ρ_M, which evidently is a solution too. The remaining 2^{NC-1} phases can be reduced by a factor 2 in the

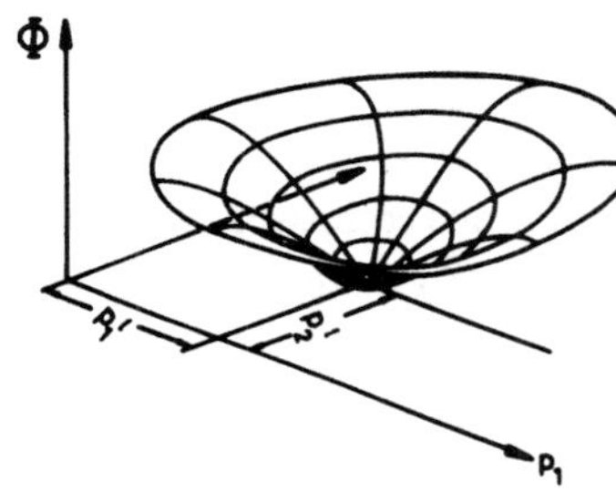

Fig. 40. The Φ-surface of a two-dimensional parameter space near a local minimum (p_1^l, p_2^l)

following way: multiplication of Fourier coefficients b_k by $(-1)^k \cdot \pm 1$ is equivalent to reflection at the origin, i.e. pass from ρ_M to $\pm\rho_{\bar{M}}$. As we assume large $\langle n \rangle$ values in (4.1), it is easy to see that a solution

$$\rho_M, \sigma_1, \sigma_2, A_1^w, A_2^w, A_3^w, A_4^w, A_5^w, A_6^w$$

is equivalent to

$$\pm\rho_{\bar{M}}, \sigma_2, \sigma_1, A_1^w, A_3^w, A_2^w, A_4^w, A_6^w, A_5^w$$

With regard to the parameters $\sigma_1, \sigma_2, A_i^w$ the Φ-surface usually exhibits from our experience only a single minimum for each set of phases, i.e. we usually find only one set of values $|b_1|, \cdots, |b_{NC}|, \sigma_1, \sigma_2, A_i^w$ for each set of phases. We therefore start the program only with 2^{NC-2} phase combinations. Take the case of 5 cos coefficients, which is equivalent to 5 distinct reflexes in $I_{exp}(b_z)$. It is easy to verify that the following 8 phases are sufficient:

+ + + + +
+ + + + −
+ + + − +
+ + + − −
+ + − + +
+ + − + −
+ + − − +
+ + − − −

In classical crystallography there is no possibility of determining the true phase combination from a single measurement of the intensity. As shown in 2.3 we have additional information compared to classical crystallography by our short Q-functions, and one can expect it to be concealed in the Φ-values of the local minima belonging to each set of phase combinations. In order to illustrate this, we now describe a computer experiment performed with the set of Eq. (4.19) and Baumeisters program[58] to solve them. The Shannon values of intensity were calculated *in advance* from a predetermined set of structural parameters and used as experimental intensity values I_m^{exp} in the following. Marquardt's procedure was started with each of the possible phase combinations and iterated until it converged to a fixed value of Φ, indicating that a local minimum had been reached. Table 2 shows that the Φ-value of the correct phase combination is several powers lower than these of every other combination and the absolute values of the structural parameters have again been determined with accuracy of a few %. We conclude that some important phase information is contained in the scattering intensity of a general paracrystalline layer lattice. We can select four solutions out of all possible solutions compared with two solutions in the case of small crystalline stacks of plane layers in Chap. 2.

In actual experiments the measurement of the experimental intensity is disturbed by a number of error sources. A further computer experiment therefore investigates roughly the error stability of our phase determination.

Table 2. Local Φ-minima for all phases of a model structure. The arrow indicates that the computer changed the phase combination during variation of parameters

Phase comb.	Φ
+ + + + +	0.25–5
+ + + + +	0.5–5
+ + + – +	↓
+ + + – –	0.14–8
+ + – + +	0.12–5
+ + – + –	0.33–5
+ + – – +	0.23–5
+ + – – –	0.77–5

Table 3. Local Φ-minima for all phases of a model structure. The model intensity was falsified in order to show experimental errors by random numbers having a rectangular distribution of width 3%, 10%, 20%, 40% of the highest intensity maximum

	3%	10%	20%	40%	40%
+ + + + +	0.32–6	0.38–6	0.42–6	0.29–6	0.45–6
+ + + + –	↑	↑	↑	↑	↑
+ + + – +	↓	↓	↓	0.32–6	↓
+ + + – –	0.84–8	0.27–7	0.16–6	0.12–6	0.24–6
+ + – + +	0.16–5	0.12–5	0.1–5	0.42–6	0.13–6
+ + – + –		0.13–5	0.13–5	↑	↑
+ + – – +	0.29–6	0.37–6	0.61–6	0.58–6	0.37–6
+ + – – –	↑	↑	↑	0.12–5	0.57–6

We used a library program which produces random numbers between –0.5 and +0.5, distributed rectangularly. We obtain a new set of *erroneous intensity values* I_m^{exp} by multiplying the random numbers by the factor (P/100) · H (H = height of highest value of I_{exp}) and adding them to the pseudo-experimental Shannon values given above. We used P = 3%, 10%, 20%, 40%, corresponding to a std. dev. of the errors of σ = 0.87%, 2.9%, 5.8%, 11.6% (% of H). For each of these sets of faulty I_m^{exp} values, the whole procedure of phase determination and iteration to the minimum was carried out; the results are given in Table 3.

The experiment with 40% was carried out with two different sets of random errors to illustrate the statistical character of the results. We clearly see from Table 3 that direct phase determination up to four phases is possible in principle, but in practice will depend upon the experimental errors. Considering Table 3 we could state that a phase is true when its Φ-value is smaller by a factor of at least 5 than the next higher Φ-value.

In experimental practice[47] we were never able up to now to find such a difference between the best Φ-value and the other Φ-values. The Φ-value alone has therefore up to now in practice not been able to select the true phase. We always had to use additional criteria. As we observed, it is a better criterion to estimate the quality of a fit by the human eye, than by the Φ-value. By inspecting the fits we can usually reject some phase combinations for their poor correspondence with the experimental intensity, the number depending on the quality of the intensity measurement. In order to find out the true solution out of the remaining phase combinations, the following methods seem to be useful:

1) Comparison of all electron density profiles evaluated from two scattering intensities having different Bragg-periods. The correct profile must have in both cases a good fit and a corresponding single membrane profile. This method is related to the swelling methods for crystalline stacks mentioned in the Introduction.
2) Comparison of all electron density profiles evaluated from two scattering intensities of biomembrane stacks of which one was treated with an electron dense contrasting substance. This method may allow to resolve the remaining ambiguity of four solutions.
3) Comparison of all electron density profiles evaluated from a scattering intensity with electron microscope pictures from the same biomembrane stack preparation[21].

We conclude with a list of possible experimental errors which prevent direct phase determination by Φ-value alone, as inferred from our experimental evidence. All should be mitigable to a tolerable level taking suitable care, during preparation and X-ray exposure of the membrane samples.

1) The sample is statistically unstable during measurement of the X-ray scattering or statistically inhomogenous within the exposed sample volume, i.e. some of the structural parameters including the Bragg-period, the flower-bunch statistics and the distribution of the number of vesicles in the stack change with time or regionally within the exposed sample volume.
2) The sample is contaminated by other paracrystalline or crystalline stacks (Lipid-Crystals for instance, which arise during preparation or measurement).
3) Macromolecules and cell-particles, if they produce a non-isotropic scattering which is not removed by the procedure described in 1.6.
4) Statistical noise in the measured scattering intensity. Improvement by simply extending the time of measurement is frequently impossible without problems 1) and 3).

5 Appendix

5.1 Proof of Incoherent Superposition of Layer and Lateral Scattering

Without loss of generality we reduce each membrane stack to only one vesicle i.e. to two membranes. The electron density of a stack with index s can then be written:

$$\rho_s = \rho_s^r + \rho_s^l * \delta\,(z - l_s)$$

where z symbolizes the Cartesian coordinate along an axis parallel to the common stacking axis of all stacks indexed by s and ρ_s^r and ρ_s^l denote the two types of three-dimensional membrane electron densities occurring in a vesicle, as shown schematically in Fig. 41.

We start with partition into layer and lateral structure:

$$\rho_s = [\rho_M\,(z_s'^r) + \Delta\rho_s^r\,(\vec{x}_s'^r)] + [\rho_{\bar{M}}\,(z_s'^l) + \Delta\rho_s^l\,(\vec{x}_s'^l)] * \delta\,(z - l_s)$$

$z_s'^r, \vec{x}_s'^r$ symbolizes the special curvilinear coordinates fitting to the membrane ρ_s^r of the s-th vesicle.

We abbreviate:

$$\rho_M\,(z_s'^r) \equiv \rho_M^{sr};\ \Delta\rho_s^r\,(\vec{x}_s'^r) \equiv \Delta\rho_{sr}$$

It then follows that:

$$\overset{2}{\tilde{\rho}}_s = \rho_s * \rho_s^- =$$

$$= \rho_M^{sr} * \rho_{\bar{M}}^{sr} + \rho_M^{sr} * \Delta\rho_{\bar{sr}} + \rho_M^{sr} * \rho_M^{sl} * \delta\,(z + l_s) + \rho_M^{sr} * \Delta\rho_{\bar{sl}} * \delta\,(z + l_s)$$

$$+\ \Delta\rho_{sr} * \rho_{\bar{M}}^{sr} + \Delta\rho_{sr} * \Delta\rho_{\bar{sr}} + \Delta\rho_{sr} * \rho_M^{sl} * \delta\,(z + l_s) + \Delta\rho_{sr} * \Delta\rho_{\bar{sl}} * \delta\,(z + l_s)$$

$$+\ \rho_M^{sl} * \rho_{\bar{M}}^{sr} * \delta\,(z - l_s) + \rho_M^{sl} * \Delta\rho_{\bar{sr}} * \delta\,(z - l_s) + \rho_M^{sl} * \rho_M^{sl} + \Delta\rho_{\bar{M}}^{sl} * \Delta\rho_{\bar{sl}}$$

$$+\ \Delta\rho_{sl} * \rho_{\bar{M}}^{sr} * \delta\,(z - l_s) + \Delta\rho_{sl} * \Delta\rho_{\bar{sr}} * \delta\,(z - l_s) + \Delta\rho_{sl} * \rho_M^{sl} + \Delta\rho_{sl} * \Delta\rho_{\bar{sl}}$$

The scattering intensity of all these stacks with identical axis are given by:

$$I_{ST}^* = \langle F\overset{2}{\tilde{\rho}}_s\rangle_s = FQ_{ST}^* = F\langle\overset{2}{\tilde{\rho}}_s\rangle_s$$

We must therefore calculate mean values of each of the above terms.

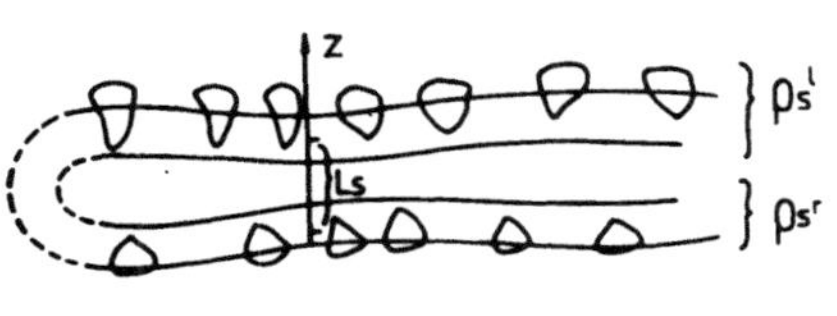

Fig. 41.

1) We first sum up all terms containing layer structure only:

$$Q_{ST} \equiv \langle \rho_M^{sr} * \rho_{\bar{M}}^{sr} + \rho_{\bar{M}}^{sl} * \rho_M^{sl} + \rho_{\bar{M}}^{sl} * \rho_{\bar{M}}^{sr} * \delta(z - l_s) + \rho_M^{sr} * \rho_M^{sl} * \delta(z + l_s)\rangle_s$$

2) We now select all terms of the form:

$$\langle \rho_M^{sr} * \Delta \rho_{sr} \rangle_s$$

Consider the set of all membranes with a particular surface form, given by a coordinate system x_c' and indexed by k. The mean value of the above can then be written:

$$\langle \rho_M^{sr} * \Delta \rho_{sr} \rangle_s = \langle\langle \rho_M (z_c') * \Delta \rho_{\bar{K}} (\vec{x}_c')\rangle_K\rangle_c$$

The inner mean value disappears, since we have assumed that there is no correlation between lateral density and membrane surface form. Therefore the whole term can be made to equal zero:

$$\langle \rho_M (z_c') * \Delta \rho_{\bar{K}} (\vec{x}_c')\rangle_K = \rho_M (z_c') * \langle \Delta \rho_{\bar{K}} (\vec{x}_c')\rangle_K = 0$$

3) The mean values of the following type also disappear

$$\langle \rho_M^{sr} * \Delta \rho_{\bar{sl}} * \delta(z + l_s)\rangle_s$$

This can be seen by selecting the set of membranes r with a particular form given by the coordinate system c_r and the set of membranes l with a coordinate system c_l and taking from all these all membrane pairs indexed by k, which have a particular distance l:

$$\begin{aligned}
&\langle \rho_M (z_s'^r) * \Delta \rho_{\bar{s}}^l (\vec{x}_s'^l) * \delta(z + l_s)\rangle_s \\
&= \langle\langle \rho_M (z_{c_r}') * \Delta \rho_K (\vec{x}_{c_l}')\rangle_K * \delta(z + l)\rangle_{l,c_l,c_r} \\
&= \langle \rho_M (z_{c_r}') * \langle \Delta \rho_K (\vec{x}_{c_l}')\rangle_K * \delta(z + l)\rangle_{l,c_l,c_r} = 0
\end{aligned}$$

Using the same argument as above.
4) In the same way, the terms of the general form:

$$\langle \Delta \rho_s^r (\vec{x}_s'^r) * \Delta \rho_s^l (\vec{x}_s'^l) * \delta(z + l_s)\rangle_s$$

must also disappear. This can be shown using the steps as above and the assumption that lateral densities of neighbouring membranes are uncorrelated. The remaining terms show the incoherent superposition of layer stack and lateral stack intensity distributions:

$$Q_{ST}^* = \langle \overset{2}{\tilde{\rho}_s}\rangle_s = Q_{ST} + \left\langle \overset{2}{\widetilde{\Delta \rho_{sr}}} + \overset{2}{\widetilde{\Delta \rho_{sl}}} \right\rangle_s \qquad (5.1)$$

$$I^*_{ST} = I_{ST} + F\left\langle \widetilde{\Delta\rho_{sr}^{2}} + \widetilde{\Delta\rho_{sl}^{2}} \right\rangle_s \tag{5.2}$$

5.2 Integrals of the Pape Procedure

5.2.1

$$\int_{-\pi+x}^{\pi} e^{i(k+l)y}\, dy \overset{k+l\neq 0}{=\!=} \frac{i(-1)^{k+l}}{k+l}\left[e^{i(k+l)x} - 1\right]$$

$$\overset{k+l=0}{=\!=} 2\pi - x$$

5.2.2

Inserting (2.18) into (2.19) we have to solve two integrals:

$$1)\ \int_0^{2\pi} \frac{i(-1)^{k+l}}{k+l}\left[e^{ikx} - e^{-ilx}\right] e^{-imx}\, dx$$

$$\overset{k+l\neq 0}{=\!=} \frac{2\pi i(-1)^{k+l}}{k+l}\left[\delta_{k,m} - \delta_{l,m}\right]$$

$$2)\ \int_0^{2\pi} e^{ikx}(2\pi - x)\, e^{-imx}\, dx = \begin{cases} 2\pi^2 & k = m \\ \dfrac{2\pi i}{k-m} & k \neq m \end{cases}$$

5.3 Effect of Vesicle Disorder on Pape's Fourier Analytical Deconvolution

We anticipate the result of Chap. 3 for the Q_{ST}-function of single vesicles and we assume for the sake of simplicity that the electron density levels of the vesicle-lumen, the vesicle outer-phase and the mean electron density of the membranes are equal to zero.

In this case Q_{ST} reduces to:

$$Q_{ST}(x \geqslant 0) = 2\,\widetilde{\rho_M^{2}} + \rho_{\bar{M}} * \rho_{\bar{M}} * \frac{1}{\sqrt{2\pi}\,\sigma_1}\, e^{-\frac{(x-d)^2}{2\sigma_1^2}} \tag{5.3}$$

Assuming a model membrane ρ_M of 100 Å thickness, this function (5.3) was evaluated for σ_1 = 2 Å, 6 Å, 9 Å, 16 Å and subsequently deconvoluted by Pape's method. Figure 42 shows the ρ_0 functions, that were found; the uppermost ρ_0 is equal to two of the original model membranes ρ_M arranged symmetrically, as it must be. At a st d.dev. $\sigma_1 = 2$ Å, Pape's method still gives the original vesicle profile and the convolution square of the result: x corresponds well with the statistically influenced Q_{ST}-function. At a larger std. dev. the distance disorder smears off the second term in (5.3). Pape's method now requires a deconvolution of $2\,\overset{2}{\tilde{\rho}}_M$. In our model ρ_M was almost symmetrical and as we see, Pape's method finds a symmetrical variation of ρ_M at a std. dev. as small as 9 Å and 16 Å, but with considerable differences between its convolution squares and the Q_{ST}-function. At std. dev. 6 Å a kind of mixture of the original vesicle profile and the symmetrized membrane profile is found.

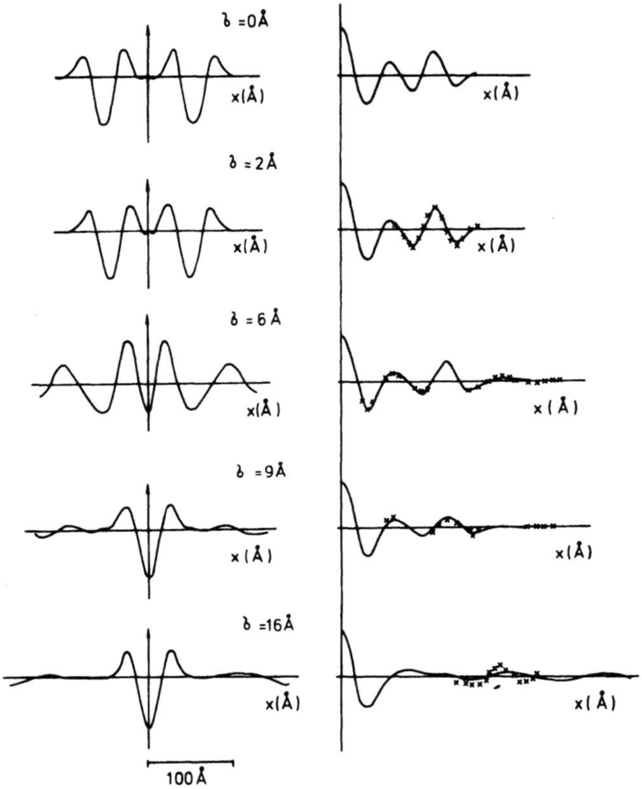

Fig. 42. Deconvolution of statistically distorted Q_0-functions with increasing st d. dev. of intermembrane distance statistics with Pape's Fourier deconvolution analysis

5.4 Derivation of the Q_{ST}-Function for the Paracrystalline Plane-Layer Lattice

Here we will show by derivation (3.10) for the simple case, that all membrane stacks are vesicles. This case will illustrate the significance of the functions *zql, zpl.* Having understood this case, the derivation of the complete expression (3.10) for arbitrary stacks is straightforward and nothing new can be learned from it. If all stacks contain only one vesicle, (3.9) reduces to (5.4):

$$\rho_s(x) = \rho_M * \delta(x - l_1) + \rho_{\overline{M}} * \delta(x - l_2) + h_I\, \sigma_1^I\, Z(|l_2 - l_1 - d|) * \delta\left(x - \frac{l_1 + l_2}{2}\right) \tag{5.4}$$

The subscript s at l_1, l_2, σ_1^I was omitted for the sake of simplicity. Figure 43 shows two types of stacks, one with $\sigma_1^I > 0$, and one with $\sigma_1^I < 0$:

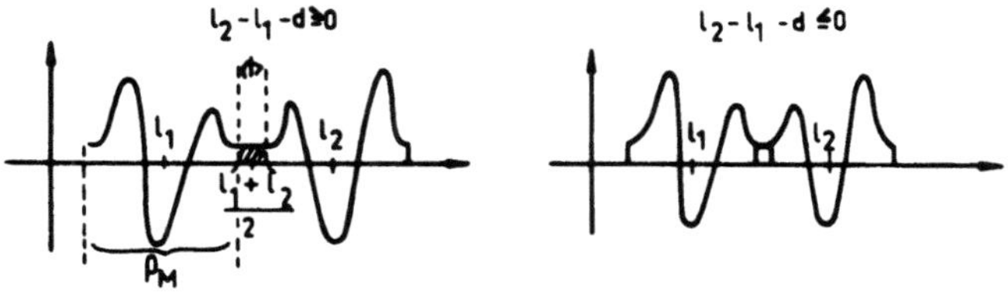

Fig. 43.

Q_{ST} is evaluated according to (3.7): $Q_{ST} = \langle \overset{2}{\widetilde{\rho_s}} \rangle_s$. Inserting (5.4):

$$\begin{aligned} Q_{ST}(x \geqslant 0) &= \langle \rho_s * \rho_{\bar{s}} \rangle \\ &= 2\, \overset{2}{\widetilde{\rho_M}} + \rho_{\overline{M}} * \rho_{\overline{M}} * \langle \delta(x - (l_2 - l_1)) \rangle_s \\ &\quad + 2\, h_I\, \rho_{\overline{M}} \left\langle \sigma_1^I \cdot Z(|l_2 - l_1 - d|) * \delta\left(x - \frac{l_2 - l_1}{2}\right) \right\rangle_s \\ &\quad + h_I^2 \langle \overset{2}{\widetilde{Z(|l_2 - l_1 - d|)}} \rangle_s \end{aligned} \tag{5.5}$$

We now calculate the single mean values of (5.5): the distance statistics (distribution of the r.v. l_1) is assumed to be a Gaussian:

$$1)\ \langle \delta(x - (l_2 - l_1)) \rangle_s = H1(x) = \frac{1}{\sqrt{2\pi}\,\sigma_1}\, e^{-\frac{(x-d)^2}{2\sigma_1^2}} \tag{5.6}$$

2) $h_I^2 \langle \overset{2}{\widetilde{Z(|l_2 - l_1 - d|)}} \rangle_s(x)$ is evidently symmetrical with regard to $x = 0$, we therefore restrict our considerations to $x \geqslant 0$:

$$\overset{2}{\widetilde{Z(a)}}(x) = \begin{cases} a - x & \text{for } x \leqslant a \\ 0 & \text{for } x > a \end{cases} \tag{5.7}$$

The mean of $\overset{2}{\widetilde{z(a)}}$ for a distribution $D(a)$ is given by:

$$\langle \overset{2}{\widetilde{Z(a)}} \rangle (x) = \int_0^\infty \overset{2}{\widetilde{z(a,x)}}\, D(a)\, da$$

In our case D (a) contains two distributions D_1 (a) and D_2 (a), D_1 (a) is the distribution of all cases with $l_2 - l_1 \geqslant d$: D_1 (a) = H1 (d + a) since $l_2 - l_1$ are distributed according to H1 (x). D_2 (a) is the distribution of all cases with $l_2 - l_1 < d_2$: $D_2(a) = H1\,(d - a)$

$$zqI\,(x) \equiv h_I^2 \langle \widehat{Z(|l_2 - l_1 - d|)}^{2} \rangle_s\,(x)$$

$$= h_I^2 \int_x^\infty (a - x)\,[H1\,(d + a) + H1\,(d - a)]\,da$$

The lower limit x of the integral accounts for the condition $a \geqslant x$ in (5.7). Inserting (5.6) and using the following well known formulae:

$$\int_x^\infty a e^{-a^2}\,da = \left[-\frac{1}{2}\,e^{-a^2}\right]_x^\infty$$

$$\phi\,(x) = \frac{1}{\sqrt{2\pi}} \int_{-\infty}^{x} e^{-\frac{a^2}{2}}\,da = 1 - \frac{1}{\sqrt{2\pi}} \int_x^\infty e^{-\frac{a^2}{2}}\,da$$

we obtain the following end result:

$$zqI\,(x) = \frac{2\,h_I^2\,\sigma_1}{\sqrt{2\pi}}\,e^{-\frac{x^2}{2\sigma_1^2}} - 2\,h_I^2 \cdot x \cdot \phi\left(\frac{-x}{\sigma_1}\right) \quad \text{for } x \geqslant 0 \tag{5.8}$$

Figure 44 shows a typical example of such a function.

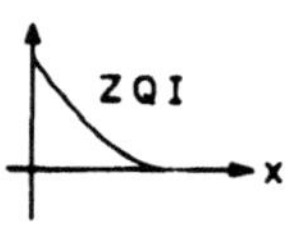

Fig. 44.

$$3)\; h_I \left\langle \sigma_1^I\; Z(|l_2 - l_1 - d|) * \delta\left(x - \frac{l_2 - l_1}{2}\right)\right\rangle_s \equiv zpI\,(x)$$

Again the summands of this mean can be divided into two classes: a) $l_2 - l_1 > d$: the summand is a positive rectangle with its left edge at d/2 and its right edge at $d/2 + l_2 - l_1 - d = l_2 - l_1 - d/2 = a + d/2$ (s. Fig. 45). b) $l_2 - l_1 < d$: the summand is a negative rectangle with its right edge at d/2 and its left edge at $d/2 - (d + l_1 - l_2) = l_2 - l_1 - d/2 = -a + d/2$ (s. Fig. 45)

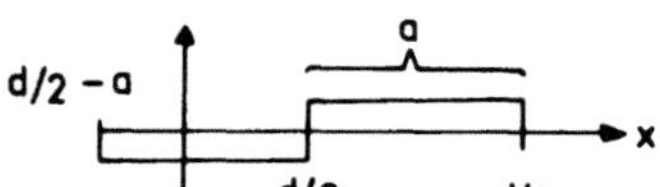

Fig. 45.

In the general formula for calculating mean values,

$$zpI\,(x) = h_I \int_0^\infty \sigma_1^I\; z\,(a, x) * \delta\left(x - \sigma_1^I \cdot \frac{a}{2} - \frac{d}{2}\right) D\,(a)\,da$$

we must insert the same distribution as above:

$D(a) = H1\ (d + a) + H1\ (d - a)$

However as can be seen from Fig. 45, for $x > d/2$ there are only contributions from positive rectangles:

$$zpI\left(x \geqslant \frac{d}{2}\right) = h_I \int_{x-d/2}^{\infty} H1\ (a + d)\ da = h_I \left(1 - \phi\left(\frac{x - d/2}{\sigma_1}\right)\right) \qquad (5.9a)$$

The lower limit x–d/2 of the integral determines that there is only a contribution from rectangles, whose right edge at a + d/2 is larger than x. For $x < d/2$ we similarly derive:

$$ZPI\left(x < \frac{d}{2}\right) = -h_I \int_{d/2-x}^{\infty} H1\left(d - a\right) da = h_I \cdot \phi\left(\frac{x - d/2}{\sigma_1}\right) \qquad (5.9b)$$

Figure 46 gives a schematic diagram of this function.

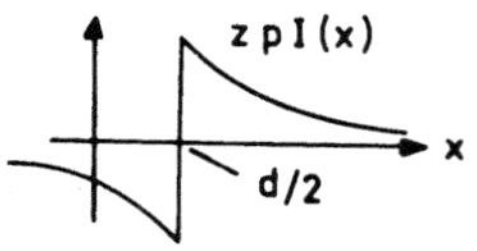

Fig. 46.

A moments reflection shows that *ZPI* is just the mean of all rectangles that have to be added to the vesicles made from constant building blocks ρ_M in order to correct for the artificially overlapping rectangles. For the special case of single vesicle stacks we have derived (3.10). Much more laborious, but straightforward is the derivation for general stacks ρ_s. Having evaluated $\overset{2}{\tilde{\rho}_s}$ one has to substitute $\sum_{i=1}^{n} \sum_{k=1}^{n} \ldots$ by introducing $j = i-k$ as in 2.1. Instead of this lengthy derivation we give a simple and graphic rule for deriving all terms of Q_{ST} for an ensemble of stacks consisting of an arbitrary number n of vesicles: Draw up a symbolic *left column* of the stack – in our case we assume two vesicles – here all building blocks are represented by their designations ρ_M and $\rho_{\bar{M}}$ respectively, and all intermembrane gaps by *zpI* and *zpE* respectively, as shown below. Draw up a symbolic *right column* where all building blocks are reflected, but all intermembrane gap symbols are unchanged relative to the left column:

ρ_M $\rho_{\bar{M}}$
zpI *zpI*
$\rho_{\bar{M}}$ ρ_M
zpE *zpE*
ρ_M $\rho_{\bar{M}}$
zpI *zpI*
$\rho_{\bar{M}}$ ρ_M

The end of the first period in the right column is marked. Now we begin with the uppermost term in the right column and form the convolution products successively with all other terms in the left column, beginning with the uppermost. At the same time we look for equivalent convolution products in the other periods of the right column and add them up. If a statistical gap is spanned, we insert the appropriate distances statistics. In this way the first term of the right column contributes:

$$2\,\rho_{\bar{M}} * \rho_M + 2\,\rho_{\bar{M}} * zpI + 2\,\rho_{\bar{M}} * \rho_{\bar{M}} * \mathrm{H1} + \rho_{\bar{M}} * zpE * \mathrm{H1}$$

$$+ \rho_{\bar{M}} * \rho_M * \mathrm{H1} * \mathrm{H2} + \rho_{\bar{M}} * zpI * \mathrm{H1} * \mathrm{H2} + \rho_{\bar{M}} * \rho_{\bar{M}} * \mathrm{H1} * \mathrm{H1} * \mathrm{H2}$$

If two statistical gaps at the same height encounter one another, we insert the appropriate zq-function. The second term of the right column therefore contributes the products: $2\,zqI + 2\,zpI * \rho_{\bar{M}} + zpI * zpE + zpI * \rho_M * \mathrm{H2} + zpI * zpI * \mathrm{H2} +$ $+ zpI * \rho_{\bar{M}} * \mathrm{H1} * \mathrm{H2}$

The third term contributes:

$$2\,\rho_M * \rho_{\bar{M}} + \rho_M * zpE + \rho_M * \rho_M * \mathrm{H2} + \rho_M * zpI * \mathrm{H2} + \rho_{\bar{M}} * \rho_{\bar{M}} * \mathrm{H1} * \mathrm{H2}$$

And the remaining convolution products are formed with the last term *zpE* of the first period of the right column:

$$zqE + zpE * \rho_M + zpE * zpI + zpE * \rho_{\bar{M}} * \mathrm{H1}$$

Substituting $n = 1$ in (3.10), we just get these terms. One might ask for some convolution products which are mainly in the region of $x \leqslant 0$ but have a *tail* into the positive range, like $zpI * \rho_M$, belonging to the second row. We have omitted them here for simplicity, but in the final programs they are inserted. This is easy, since some of the terms, given in (3.10) have *tails* ranging into $x \leqslant 0$. The tails that are omitted, are identical with these latter tails after reflexion at the origin.

5.5

1) One-sphere-cap density function $\eta\,(\alpha, z')$ of pillars having angle α and height z'.

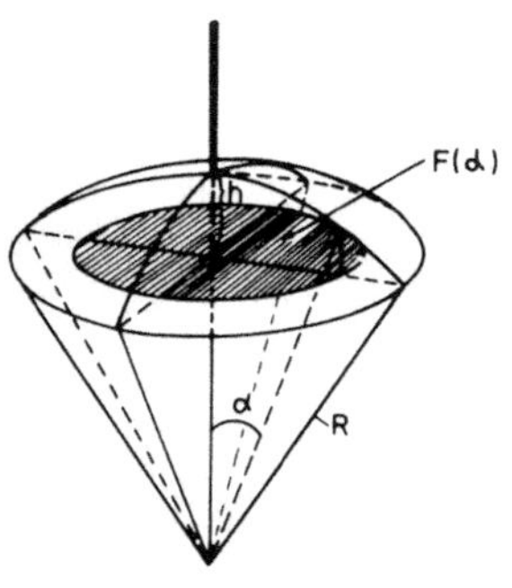

Fig. 47.

The following geometrical relationships follow from Fig. 47:

$$z' = R(\cos\alpha - 1) + h$$

$$\cos\alpha = \frac{1}{R}(z' - h + R)$$

$$F(\alpha) = R^2 \sin^2\alpha \cdot \pi$$

The number of pillars within an interval $(\alpha, d\alpha)$ is proportional to $\frac{dF(\alpha)}{d\alpha}\, d\alpha$. The projection integral can thus be expressed as an integral:

$$[P_z\rho_0](z) = \int d\alpha\, \rho_0([z - R\{\cos\alpha - 1\} - h]\cos\alpha)\, 2\pi R^2 \sin\alpha\cos\alpha$$

But the same can be expressed as a z' integral:

$$[P_z\rho_s](z) = \int dz'\, \rho_0\left([z - z']\frac{z' - h + R}{R}\right)\frac{dF(z')}{dz'}$$

$$F(z') = 2\pi\left(z'h - z'R + hR - \frac{1}{2}z'^2 - \frac{1}{2}h^2\right)$$

It follows:

$$[P_z\rho_0](z) = \int dz'\, \rho_0\left([z - z']\cdot\frac{1}{R}[z' - h + R]\right)\cdot 2\pi(h - R - z')$$

Both integrals can be combined formally:

$$[P_z\rho_0](z) = \int d\alpha \int dz'\, \rho_0([z - z']\cos\alpha)\,\delta(R(\cos\alpha - 1) + h - z')\cdot 2\pi R^2 \sin\alpha\cos\alpha$$

This can be easily proved by carrying out either of the two integrations and using the rule for integration of δ-functions (2.7). Comparing the result above with (4.7), we conclude that for a spherical cap-segment of radius R and height h we have:

$$\eta_R^h(z', \alpha) = \text{const}\cdot\delta(R[\cos\alpha - 1] + h - z')\, 2\pi R^2 \sin\alpha\cos\alpha \qquad (5.10)$$

2) Density function of an ensemble of spherical caps, showing Gaussian distributions of R and h:

$$\eta_{\langle R\rangle}^{\langle h\rangle}(z', \alpha) \equiv \text{const.} \iint \eta_R^h(z', \alpha)\cdot e^{-\frac{(R - \langle R\rangle)^2}{2\sigma_R^2}}\cdot e^{-\frac{(h - \langle h\rangle)^2}{2\sigma_h^2}}\, dR\, dh$$

$$= 2\pi\sin\alpha\cos\alpha\cdot C\cdot\int dR\cdot R^2 \exp\left(-\left[\frac{(R - \langle \mathbf{R}\rangle)^2}{2\sigma_R^2} + \frac{(z' - R(\cos\alpha - 1) - \langle \mathbf{h}\rangle)^2}{2\sigma_h^2}\right]\right)$$

Introducing:

$$A \equiv \frac{1}{2\,\sigma_R^2\,\sigma_h^2}\,[\sigma_h^2 + (\cos\alpha - 1)^2\,\sigma_R^2]$$

$$B \equiv \frac{1}{2\,\sigma_R^2\,\sigma_h^2}\,[2\,(\cos\alpha - 1)\,\langle \mathbf{h}\rangle\,\sigma_R^2 - 2\,\langle \mathbf{R}\rangle\,\sigma_h^2 - 2\,z'\,(\cos\alpha - 1)\,\sigma_R^2]$$

$$C \equiv \frac{1}{2\,\sigma_R^2\,\sigma_h^2}\,[\langle \mathbf{R}\rangle^2\,\sigma_h^2 + z'^2\,\sigma_R^2 + \langle \mathbf{h}\rangle^2\,\sigma_R^2 - 2\,z'\,\langle \mathbf{h}\rangle\,\sigma_R^2]$$

we can transform the expression above into:

$$= \exp\left[\frac{B^2}{4\,A} - C\right] \int dR\,R^2\,\exp\left(-A\left[R + \frac{B}{2\,A}\right]^2\right)$$

substituting $r = R + \dfrac{B}{2A}$ and using the well-known formulae:

$$\int_{-\infty}^{+\infty} dx\,e^{-Ax^2} = \sqrt{\frac{\pi}{A}}\,; \quad \int_{-\infty}^{+\infty} dx\,x^2\,e^{-Ax^2} = \frac{1}{2\,A}\sqrt{\frac{\pi}{A}}$$

$$\int_{-\infty}^{+\infty} dx \cdot x \cdot e^{-Ax^2} = 0$$

we find the following result:

$$\eta_{\langle R\rangle}^{\langle h\rangle}(z', \alpha) = \text{Const} \cdot 2\,\pi \cdot \sin\alpha \cdot \cos\alpha\,\exp\left[\frac{B^2}{4\,A} - C\right] \cdot \sqrt{\frac{\pi}{A}}\left[\frac{1}{2A} + \frac{B^2}{4\,A^2}\right] \qquad (5.11)$$

The constant was determined by computer, to meet the norm:

$$\iint dz'\,d\alpha\,\eta_{\langle \mathbf{R}\rangle}^{\langle \mathbf{h}\rangle}(z', \alpha)\,\frac{1}{\cos\alpha} = 1 \qquad (5.12)$$

The factor $1/\cos(\alpha)$ is a result of the volume difference of pillars, having different α, so (5.12) is a norm for the total volume, that contributes to the scattering intensity. Figure 48 shows a computer plot of a typical function (5.11). The function was symmetrized to negative α in order to show its both *walls.* The parameters of Fig. 28 were chosen in the following way:

$\langle \mathbf{h}\rangle =$	2	10	20	30	40 Å	$\sigma_h = 7$ Å
$\langle \mathbf{R}\rangle =$	8000	1600	800	500	400 Å	$\sigma_R = 15$ Å

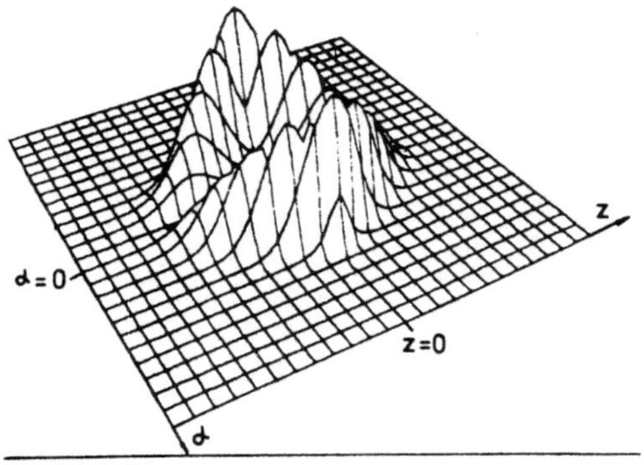

Fig. 48. The distribution $\eta_{\langle R\rangle}^{\langle h\rangle}(z, \alpha)$ for $\langle \mathbf{R}\rangle = 500$ Å, $\sigma_R = 10$ Å, $\langle \mathbf{h}\rangle = 50$ Å, $\sigma_h = 10$ Å. The mirror image at the $\alpha = 0$ axis was added, to show the "other side" of the curved ridge

5.6 Derivation of a System of Equations Equivalent to (4.16)

Representation of the membrane profile:

$$\rho_M(z + d/2) = \begin{cases} \sum_{k=-NC}^{NC} c_k\, e^{2\pi i \frac{k}{2d}\cdot z} & -d \leqslant z \leqslant 0 \\ 0 & \text{otherwise} \end{cases} \tag{5.13}$$

with $c_k = c_{-k}$ and real by symmetry. Figure 49 shows the series above schematically.

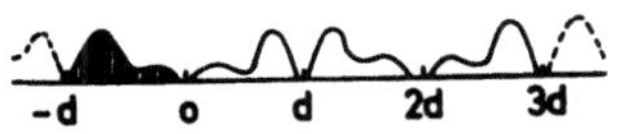

Fig. 49.

c_k are Shannon-values of the following structure:

$$\rho_0(z) \equiv \rho_M\left(z + \frac{d}{2}\right) + \rho_M\left(-z + \frac{d}{2}\right)$$

The experimental Q-function $Q_{exp} = F\, I_{exp}$ disappears outside an interval $\left[\frac{-Nd}{2}, \frac{Nd}{2}\right]$ with a suitable chosen number N. We therefore represent I_{exp} by the set of Shannon-values:

$$I_{exp}\left(\frac{m}{Nd}\right), \; m = 1, \ldots$$

It will prove useful to convert the coefficients c_k of (5.13) into the Shannon-values d_l at $b_l = 1/(Nd)$ of the Fourier transform of ρ_M; they are equivalent to the coefficients of a series with periodicity interval $\left[\frac{-Nd}{2}, \frac{Nd}{2}\right]$ and represent ρ_M once in each period, as shown in Fig. 50.

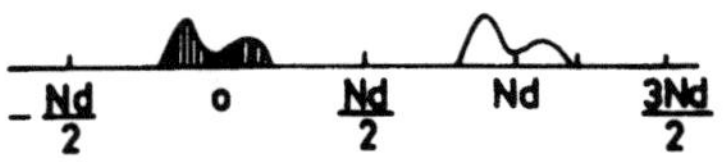

Fig. 50.

$$\rho_M(z) = \begin{cases} \sum\limits_e d_e\, e^{2\pi i \frac{l}{Nd} z} & \frac{-Nd}{2} \leqslant z \leqslant \frac{Nd}{2} \\ 0 & \text{otherwise} \end{cases}$$

$$d_e = \frac{1}{Nd} \int\limits_{-d/2}^{d/2} \left\{ \sum_{k=1}^{NC} c_k\, e^{2\pi i \frac{k}{2d}\left(z - \frac{d}{2}\right)} \right\} e^{-2\pi i \frac{l}{Nd} z}$$

After a straightforward calculation:

$$d_e = \sum_{k \neq \frac{2l}{N}}^{Ne} c_k\, (-i)^k \frac{2}{\pi\,(kN - 2l)} \sin\left\{\frac{\pi}{2}\left(k - \frac{2l}{N}\right)\right\}$$

$$\text{if } k = \frac{2l}{N}: + \frac{1}{N}\, C_{\frac{2e}{N}}\, (-i)^{\frac{2l}{N}}$$

Furthermore we need the Fourier-coefficients from *zqI, zqE, zpI, zpE* when expanded into the same sort of series:

$$1)\ ZQI\,(l) \equiv \frac{1}{(Nd)^2} \int\limits_{-\frac{Nd}{2}}^{\frac{Nd}{2}} zqI\,(z)\, e^{-2\pi i \frac{l}{Nd} z}\, dz$$

using the symmetry of *zqI* and (5.8):

$$ZQI\,(l) = \frac{h_I^2}{2\,\pi^2 l^2} \left(1 - \exp\left(\frac{-2\,\pi^2\,\sigma_1^2\,l^2}{N^2\,d^2}\right)\right) \tag{5.14}$$

We have used[56]:

$$\int\limits_s^{\infty} \sin(xt)\, \mathrm{Erf\,c}\,(yt)\, dt = \frac{1}{x}\left[1 - \exp\left(\frac{-x^2}{4y^2}\right)\right] \qquad x,\ \mathbf{Re}\, y > 0 \tag{5.15}$$

deriving this formula with respect to x and using

$$\phi(t) = \frac{1}{2}\, \mathrm{Erfc}\left(\frac{-t}{\sqrt{2}}\right)$$

we find (5.14). *ZQE* (1) is the same with h_I, σ_1 substituted by h_E, σ_2.

$$2)\ ZPI\,(l) = \frac{1}{Nd} \int_{-\frac{Nd}{2}}^{\frac{Nd}{2}} zpl\,(z)\, e^{-2\pi i \frac{l}{Nd} z}\, dz$$

Using the antisymmetry of *zpl* with respect to z = d/2 we find:

$$= e^{-\frac{\pi i l}{N}} \cdot \frac{2i}{Nd} \int_0^{Nd/2} zpl\,(z + d/2) \sin\left(\frac{2\pi l}{Nd} z\right) dz$$

Inserting (5.9) and using (5.15), we find the result:

$$ZPI\,(l) = \frac{-i\, h_l}{2\pi l}\, e^{-\frac{i\pi l}{N}} \left[1 - \exp\left(\frac{-2\pi^2 l^2 \sigma_1^2}{N^2 d^2}\right)\right] \tag{5.16}$$

and an analogous formula for *ZPE* (l).

In the same way we obtain the Fourier coefficients (or Shannon values) of H1 (z + d) and H2 (z + d):

$$H1\,(l) \equiv e^{-\frac{2\pi^2 l^2 \sigma_1^2}{N^2 d^2}}\ ; H2\,(l) \equiv e^{-\frac{2\pi^2 l^2 \sigma_2^2}{N^2 d^2}} \tag{5.17}$$

With the d_l of above, the convolution products of (4.16) are given by:

$$\underset{\sim}{\overset{2}{\rho_M}} \equiv \begin{cases} \sum_l |d_l|^2\, e^{2\pi i \frac{l}{Nd} z} & -\frac{Nd}{2}\, z \leqslant \frac{Nd}{2} \\ 0 & \text{otherwise} \end{cases} \tag{5.18}$$

$$\rho_M * \rho_M = \begin{cases} \sum_l (d_l)^2\, e^{2\pi i \frac{l}{Nd} z} & -\frac{Nd}{2} \leqslant z \leqslant \frac{Nd}{2} \\ 0 & \text{otherwise} \end{cases}$$

$\rho_{\bar M} * \rho_{\bar M}$ as $\rho_M * \rho_M$, but with $(d_l^*)^2$ instead of d_l^2. In the same way the functions c_1, c_2, c_3 of (4.16) can be expressed:

$$c_1\,(g) = \sum_l \{2\langle n-g\rangle |d_l|^2 + 2\langle n-g\rangle d_l^* \cdot ZPI\,(l)$$
$$+ 2\langle n-g-1\rangle \cdot d_l \cdot ZPE\,(l) + 2\langle n-g-1\rangle ZPI\,(l) \cdot ZPE\,(l)\} \cdot e^{2\pi i \frac{l}{Nd} z} \cdot \frac{\delta\,(R)}{\pi R}$$

with similar formulas for c_2 and c_3, or abbreviated with obvious definitions:

$$c_1(g) = \sum_l C_1(l, g)\, e^{2\pi i \frac{l}{Nd} z} \cdot \frac{\delta(R)}{\pi R} \tag{5.19}$$

$$C_0(l) \equiv \langle n\rangle \cdot ZQI(l) + \langle n-1\rangle \cdot ZQE(l) - 2\langle n\rangle \cdot |d_l|^2$$

with similar definitions for $C_2(l, g)$ and $C_3(l, g)$. Substituting these formulae into (4.16), we obtain a new function $\mathbf{Q}_{ST}^{per}(R, z)$ as a function of parameters:

$d_l, \sigma_1, \sigma_2, A_i^w$

$\mathbf{Q}_{ST}^{per}$ is a periodic recording of $\mathbf{Q}_{ST}$ with interval $\left[-\frac{Nd}{2}, \frac{Nd}{2}\right]$:

$$\mathbf{Q}_{ST} = \begin{cases} \mathbf{Q}_{ST}^{per} & -\frac{Nd}{2} \leqslant z \leqslant \frac{Nd}{2} \\ 0 & \text{otherwise} \end{cases}$$

$\mathbf{I}_{ST}^{per} = F\mathbf{Q}_{ST}^{per}$ therefore resembles $F\mathbf{Q}_{ST}$ in the Shannon-points $b_l = l/Nd$. As can be seen from (4.16b), we now have to evaluate the Fourier transform of $\mathbf{Q}_{ST}$. Using Shannon's theorem, we are able to reconstruct I_{ST} by convolution of $\mathbf{I}_{ST}^{per}$ with

$\frac{\sin(\pi N d\, b_z)}{\pi b_z}$[52].

First step: $\mathbf{I}_{ST}^{per} = F\mathbf{Q}_{ST}^{per}$. Using the convolution theorem we obtain from (4.16):

$$\begin{aligned} \mathbf{I}_{ST}^{per} &= F\{B_0 * H_I^w + [B_1 * H_1^w + \cdots + B_6 * H_6^w] \\ &\qquad + [B_1 * H_1^w + \cdots + B_6 * H_6^w]^- \} \\ &= B_0 \cdot H_0^w + 2\,\mathbf{Re}\,\{B_1 \cdot H_1^w + \cdots + B_6 \cdot H_6^w\} \end{aligned} \tag{5.20}$$

$$B_1(b_R, b_z) \equiv F c_1(0) = \sum_l C_1(l, 0) \cdot \delta\left(b_z - \frac{l}{Nd}\right) \cdot 1(b_R)$$

With obvious definitions for $B_0, B_2, \cdots, B_6$.

Furthermore:

$$H_i^w(b_R, b_z) = F H_i^w(R, z)$$

$$= \int d_R \int d_z \left\{ \frac{1}{\sqrt{2\pi}\, A_i^w R} \cdot e^{-\frac{z^2}{2(A_i^w R)^2}} \right\} J_0(2\pi b_R \cdot R) \cdot R \cdot e^{-2\pi i b_z \cdot z}$$

We begin with the z-integration, after which we substitute

$$R = \bar{R}\, \frac{1}{\sqrt{2\pi}\, A_i^w b_z}\,; \quad b_R = \bar{b}_R \sqrt{2\pi}\, A_i^w b_z$$

and using the Hankel-Transform formula[57]:

$$2\pi \int_0^\infty e^{-\pi \bar{R}^2} J_0 (2\pi \bar{b}_R \cdot \bar{R}) \cdot \bar{R} = e^{-\pi \bar{b}_R^2}$$

we obtain

$$H_i^w (b_R, b_z) = \frac{1}{(2\pi A_i^w \cdot b_z)^2} e^{-\frac{b_R^2}{2(A_i^w b_z)^2}} \tag{5.21}$$

Second step: reconstitution of I_{ST} from its Shannon-points:

$$I_{ST} (b_R, b_z) = I_{ST}^{per} (b_R, b_z) * \frac{\sin(\pi N d\, b_z)}{\pi b_z}$$

Having inserted (5.20), use the relation[49c]:

$$f(x) \cdot \delta (z - y) = f(y) \cdot \delta (z - y)$$

i.e. we substitute $b_z = 1/Nd$ for all b_z-arguments in (5.21). The convolution integral then reduces to:

$$\int \delta \left(b_z' - \frac{1}{Nd}\right) \frac{\sin(\pi N d (b_z - b_z'))}{\pi (b_z - b_z')} db_z' = \frac{\sin\left(\pi N d\left(b_z - \frac{1}{Nd}\right)\right)}{\pi\left(b_z - \frac{1}{Nd}\right)}$$

The following then applies:

$$\begin{aligned} I_{ST} (b_R, b_z) = \sum_l \Big[& C_0 (l) \cdot H_1^w \left(b_R, \frac{1}{Nd}\right) + 2\,\mathbf{Re} \Big\{ C_1 (l, 0) \cdot H_1^w \left(b_R, \frac{1}{Nd}\right) \\ & + C_2 (l, 0) \cdot H_2^w \left(b_R, \frac{1}{Nd}\right) + C_3 (l, 0) \cdot H_3^w \left(b_R, \frac{1}{Nd}\right) \\ & + C_1 (l, 1) \cdot H1 (l) \cdot H2 (l) \cdot H_4^w \left(b_R, \frac{1}{Nd}\right) \\ & + C_2 (l, 1) \cdot H1 (l) \cdot H2 (l) \cdot H_5^w \left(b_R, \frac{1}{Nd}\right) \\ & + C_3 (l, 1) \cdot H1 (l) \cdot H2 (l) \cdot H_6^w \left(b_R, \frac{1}{Nd}\right) \Big\} \Big] \cdot \frac{\sin\left(\pi N d\left(b_z - \frac{1}{Nd}\right)\right)}{\pi\left(b_z - \frac{1}{Nd}\right)} \end{aligned} \tag{5.22}$$

The integral (4.16b), which must now be evaluated, is solved conveniently in spherical coordinates:

$$b_R = b \sin \alpha \qquad b_z = b \cos \alpha$$

$$I_{exp}(b) = \int_0^{\pi} d\alpha\, h(\alpha) \sin\alpha\, I_{ST}(b \sin\alpha, b\cos\alpha)$$

However we require only the Shannon values $I_m^{Mod} \equiv I_{exp}\left(\frac{m}{Nd}\right)$:

$$I_m^{Mod} = \int_0^{\pi} d\alpha\, h(\alpha) \sin\alpha\, I_{ST}\left(\frac{m}{Nd}\sin\alpha, \frac{m}{Nd}\cos\alpha\right)$$

$$\begin{aligned} I_m^{Mod} = \sum_1 \, [&C_0(1) \cdot J_m^1(A_1^w) + \\ &+ 2\,\mathbf{Re}\,\{C_1(1,0)\cdot J_m^1(A_1^w) + C_2(1,0)\cdot J_m^1(A_2^w) \\ &+ C_3(1,0)\,J_m^1(A_3^w) + C_1(1,1)\cdot H1(1)\cdot H2(1)\cdot J_m^1(A_4^w) \\ &+ C_2(1,1)\cdot H1(1)\cdot H2(1)\cdot J_m^1(A_5^w) \\ &+ C_3(1,1)\cdot H1(1)\cdot H2(1)\cdot J_m^1(A_6^w)\}\,] \end{aligned} \qquad (5.23)$$

$$J_m^1(A) \equiv Nd \int_0^{\pi} d\alpha\, h(\alpha) \sin\alpha\, H_i^w\left(\frac{m}{Nd}\sin\alpha, \frac{1}{Nd}\right) \frac{\sin(\pi(m\cos\alpha - 1))}{\pi(m\cos\alpha - 1)}$$

The numbers A_i^w are the only structural unknowns on which these functions J_m^1 depend. An analytical expression for these integrals could not be found. Computation of all these integrals within each iteration of Marquardts program needs a great deal computing time.

Fortunately, the region into which A_i^w are varied can be considerably restricted:

Lower bound: in the case of our biomembrane stacks $\frac{Nd}{2} \cong 200$ Å. The most prominent Fourier waves of ρ_M on the other hand have half-wavelengths of about 20 Å. Therefore I_{ST} can reasonably be approximated by the plane membrane approximation, when A_i^w are smaller than 0.1. In this region I_m^{Mod} does not change very much with A_i^w, as was confirmed by computer experiments.

Upper bound: at $A_i^w \geqslant 0.35$ the convolution products smear off at a distance of about 80 Å from the z-axis and their contributions can be neglected usually.

The interval of interest $0.1 \leqslant A_i^w \leqslant 0.35$ could reasonably be covered by developing the J_m^1-functions in three Taylor series, each of the order two, ranging in three intervals:

$$[0.08 \leqslant A_i^w \leqslant 0.15]; [0.15 < A_i^w \leqslant 0.25]; [0.25 < A_i^w \leqslant 0.35]$$

The convergence to the correct values of J_m^1 of these series was tested. Application of the program to a particular small-angle scattering intensity of biomembrane stacks starts with measuring $h(\alpha)$ – we adopt a Gaussian model – and calculating the Taylor coefficients of all the functions $J_m^1(A)$ for all pairs $1 \leqslant m$, $1 \leqslant 32$ which is sufficient for all our intensities up to now. These Taylor coefficients are later read into Marquardts program, which uses them to evaluate J^{1}_{m} (A_i^w) for the temporary parameter values.

Acknowledgements. We should like to express our gratitude for fruitful discussions to M. Frei, J. Funk, M. Gborzoe, K. Haag, L. Jaenicke, P. Mahler, C. Müller, W. Müller-Klieser, E. H. Pape, H. Reidel, S. Stange and D. Walter. We are especially grateful to Mr. N. Hodapp with whom we consulted closely throughout these investigations. All the computer calculations were done at the computer centre of the University of Freiburg (Rechenzentrum der Universität Freiburg), where the staff always gave us friendly assistance. Mr. L. S. Davies kindly corrected the English draft of this article. Last but not least, our thanks are to Mrs. El-Deeb, Mr. E. Waidele and especially Mrs. G. Heppeler for their excellent preparation of the figures.

6 References

1. Zernicke, F., Prins, J. A.: Die Beugung von Röntgenstrahlen in Flüssigkeiten als Effekt der Molekülanordnung. Z. f. Physik *41,* 184 (1927)
2. Kratky, O.: Die Struktur des flüssigen Quecksilbers. Physikal. Zeitschr. *34,* 482 (1933)
3. Levine, Y. K., Wilkins, M. H. F.: Structure of oriented Lipid Bilayers. Nature *230,* 69 (1971)
4. Engelmann, D. M.: Lipid bilayer Structure in the Membrane of Mycoplasma laidlawii. J. Mol. Biol. *58,* 153 (1971)
5. Blaurock, A. E.: X-Ray Diffraction Pattern from a Bilayer with Protein outside. Biophysical J. *13,* 281 (1973)
6. Blaurock, A. E.: The Structure of a Lipid-Cyt. C Membrane. Biophysical J. *13,* 290 (1973)
7. Lee, R. E., Worthington, C. R., Glew, R. H.: The Bilayer Nature of Deposits Occurring in Gaucher's Disease. Arch. of Biochem. & Biophys. *159,* 259 (1973)
8. Franks, N. P.: Structural Analysis of Hydrated Egg-Lecithin and Cholesterol Bilayers. I. X-Ray Diffraction. J. Mol. Biol. *100,* 345 (1976)
9. Worcester, D. L., Franks, N. P.: Structural Analysis of Hydrated Egg Lecithin and Cholesterol Bilayers. II. Neutron Diffraction. J. Mol. Biol. *359,* 100 (1976)
10. Podo, F., Cain, J. E., Blasie, J. K.: Structural and Dynamical Studies of Mixed Chlorophyll/Phosphatidylcholine Bilayers via X-Ray Diffraction, Absorption Polarization Spectroscopy and Nuclear Magnetic Resonance. BBA *419,* 19 (1976)
11. Khare, R. S., Worthington, C. R.: An X-Ray Diffraction Study of Sphingomyelin-Cholesterol Interaction in Oriented Bilayers. Mol. Cryst. Liq. Cryst. *38,* 195 (1977)
12. Casper, D. L. D., Kirschner, D. A.: Myelin Membrane Structure at 10 Å Resolution. Nature *231,* 46 (1971)
13. McIntosh, Worthington, C. R.: Direct Determination of the Lamellar Structure of Peripheral Nerve Structure at low Resolution (17 Å). Biophysical J. *14,* 363 (1974)
14. Worthington, C. R., McIntosh, T. J.: Direct Determination of the Lamellar Structure of Peripheral Nerve Myelin at moderate Resolution. Biophysical J. *14,* 703 (1974)
15. Blaurock, A. E., Wilkins, M. H. F.: Structure of Frog-Photoreceptor Membranes. Nature *223,* 906 (1969)
16. Chabre, M.: X-Ray Diffraction Studies of Retinal Rods. I. Structure of the Disc Membrane, Effect of Illumination. II. Light Effect on the Osmotic Properties. BBA *382,* 322 (1975)
17. Makowski, L., Caspar, D. L. D., Phillips, W. C., Goodenough, D. A.: Gap Junction Structures. II. Analysis of the X-Ray Diffraction Data. The J. of Cell Biol. *74,* 629 (1977)
18. Blaurock, A. E., Stoeckenius, W., Oesterhelt, D., Scherphof, G. L.: Structure of the Cell Envelope of Halobacterium Halobium. The J. of Cell Biol. *71.* 1 (1976)

19. Blaurock, A. E., Wober, W.: Structure of the Wall of Halobacterium halobium Gas Vesicles. J. Mol. Biol. *106*, 871 (1976)
20. Blaurock, A. E., King, G. I.: Asymetric Structure of the Purple Membrane. Science *196*, 1101 (1977)
21. Sadler, D. M., Lefort-Tran, M., Pouphile, M.: Structure of Photosynthetic Membranes of Euglena using X-Ray Diffraction. BBA *298*, 620 (1973)
22. Sayre, D.: Some Implications of a Theorem due to Shannon. Acta Cryst. *5*, 843 (1952)
23. Moody, M. F.: Structure Determination of Membranes in Swollen Lamellar Systems. Biophys. J. *14*, 697 (1974)
24. Worthington, C. R., King, G. I., McIntosh, T. J.: Direct Structure Determination of Multilayered Membrane-Type Systems Which Contain Fluid Layers. Biophys. J. *13*, 480 (1973)
25. Lesslauer, W., Cain, J., Blasie, J. K.: On the Location of 1-Anilino-8-Naphtalene-Sulfonate in Lipid Model Systems. An X-Ray Diffraction Study. BBA *241*, 547 (1971)
26. Lesslauer, W.: X-Ray Diffraction from Fatty-Acid Multilayers. Significance of Intensity Data in Low-Angle Diffraction. Acta Cryst. *B 30*, 1927 (1974)
27. Lesslauer, W.: X-Ray Diffraction from Fatty-Acid Multilayers. Angular Width of Reflections from Systems with Few Unit Cells. Acta Cryst. *B 30*, 1932 (1974)
28. Mahler, P. H.: Direkte Bestimmung des Elektronendichteprofils von Cadmium-di-Stearat unter besonderer Berücksichtigung der Messfehlerempfindlichkeit des Auswerteverfahrens. Diplomarbeit, Institut für Biophysik & Strahlenbiol. der Universität Freiburg i. Br. (1977)
29. Blaurock, A. E., Stoeckenius, W.: Structure of the Purple Membrane. Nature *233*, 153(1971)
30. Porod, G.: Die Röntgenkleinwinkelstreuung von dicht gepackten kolloiden Systemen. Kolloid Zeitschrift *124*, 83 (1951)
31. Ewald, P. P.: X-Ray Diffraction by Finite and Imperfect Crystal Lattices. Proc. Phys. Soc. (London) *52*, 167 (1940)
32. Hosemann, R., Bagchi, S. N.: Direct analysis of diffraction by matter. Amsterdam: North-Holland Publishing Company, 1962.
a) Chap. I; b) Chap. II § 1; c) Chap. IV § 2; d) Chap. VII § 9
33. Pape, E. H.: A New Deconvolution Method for Evaluating Electron Density Distributions from Small Angle Scattering Diagrams. Biophys. J. *14*, 284–294 (1974)
34. Kreutz, W., Menke, W.: Strukturuntersuchungen an Plastiden. III. Z. f. Naturforschung *17b*, 675 (1962)
35. Kreutz, W.: Strukturuntersuchungen an Plastiden. VI. Z. f. Naturforschung *19 b*, 441(1964)
36. Kreutz, W.: X-Ray Structure Research on the Photosynthetic Membrane, in: Advances in Botanical Research. R. D. Preston (ed.). London and New York: Academic Press, Vol. 3, p. 53–168 (1970)
37. Schwartz, S., Cain, J. E., Dratz, E. A., Blasie, J. K.: An Analysis of Lamellar X-Ray Diffraction from Disordered Membrane Multilayers with Application to Data from Retinal Rod outer Segments. Biophys. J. *15*, 1201–1233 (1975)
38. Pape, E. H., Klott, K., Kreutz, W.: The Determination of the Electron Density Profile of the Human Erythrocyte Ghost Membrane by Small-Angle X-Ray Diffraction. Biophys. J. *19*, 141–161 (1977)
39. Kreutz, W., Pape, E. H.: Evaluation of X-ray diffraction patterns of biomembrane systems, in: Summer institute on the physics of biological membranes. Colbow, K. (ed.). Vancouver, B. C., Canada: Simon Fraser University (1974)
40. Blaurock, A. E., Nelander, J. C.: Disorder in Nerve Myelin: Analysis of the Diffuse X-Ray Scattering. J. Mol. Biol. *103*, 421 (1976)
41. Nelander, J. C., Blaurock, A. E.: Disorder in Nerve Myelin: Phasing the Higher Order Reflexions by Means of the Diffuse Scatter. J. Mol. Biol. *118*, 497 (1978)
42. Bonart, R.: Parakristalline Schichtgitter in verstrickten Linearpolymeren. I. Die Querstruktur. Kolloid-Zeitschrift & Zeitschrift für Polymere *194*, 97 (1964)
43. Gbordzoe, M. K., Kreutz, W.: Direct X-Ray Determination of the Electron-Density Profile of the Nerve Myelin Membrane, with Paracrystalline Lattice Distortions Taken into Account. J. Appl. Cryst. *11*, 489 (1978)

44. Herbette, L., Marquardt, J., Scarpa, A., Blasie, J. K.: A Direct Analysis of Lamellar X-Ray Diffraction from Hydrated Oriented Multilayers of Fully Functional Sarcoplasmic Reticulum. Biophys. J. *20,* 245 (1977)
45. Cain, J. E.: Lattice and Substitution Disorder in Oriented Membrane Systems, Structure of the Chromatophore Membrane by X-Ray Diffraction. Federation Proceedings *33,* 1241 (1974)
46. Blasie, J. K., Erecinska, M., Samuels, S., Leigh, J. S.: The Structure of a Cytochrome Oxidase-Lipid Model Membrane. BBA *501,* 33 (1978)
47. Hodapp, N., Kreutz, W.: To be published
48. Müller-Klieser, Kreutz, W.: On the Cross-Section Structure of the Mitochondrial Cristae-Membrane as Revealed by X-Ray Diffraction. Z. f. Naturforschung *31c*, 612 (1976)
49. Papoulis, A.: Systems and transforms with applications in optics. Toronto-New York: MacGraw-Hill Book Company 1968.
a) Chap. II § 1; b) Chap. II § 3; c) Chap. II § 2; d) Chap. II § 2
50. Papoulis, A.: Probability, random variables and stochastic processes. Toronto-New York: McGraw-Hill Book Company 1965, Chaps. III and IV
51. Deas, H. D.: The Diffraction of X-Rays by a Random Assemblage of Molecules Having Partial Alignment. Acta Cryst. *5,* 542 (1952)
a) Sec. 4
52. Papoulis, A.: The fourier integral and its applications. Toronto, New York: McGraw-Hill Book Company 1962
53. Marquardt, D. W.: An Algorithm for Least-Squares Estimation of Nonlinear Parameters. Journal Soc. Indust. and Applied Mathematics *2,* 431 (1963)
54. Bevington, P. R.: Data reduction and error analysis for the physical sciences. Toronto, New York: McGraw-Hill Book Company 1969, Chap. 11
55. Blaurock, A. E.: Disorder is Characteristic of Nerve Myelin. BBA *510,* 11 (1978)
56. Magnus, W., Oberhettinger, F., Soni, R. P.: Formulas and theorems for the special functions of mathematical physics. Berlin, Heidelberg, New York: Springer 1966, Chap. 9.2.3.
57. Bracewell, R.: The fourier Transform and its applications. Toronto, New York: McGraw-Hill Book Company 1965, Chap. 12
58. Marquardt, D. W., Baumeister, T., Sheldon, J. A., Stanley, R. M.: Least Squares Estimation of Nonlinear Parameters. Engineering Department E. I. du Pont de Nemours & Co., Inc. Wilmington, Delaware

Received September 18, 1978
H.-J. Cantow (editor)

Author Index Volumes 1-30

GPSR Compliance

The European Union's (EU) General Product Safety Regulation (GPSR) is a set of rules that requires consumer products to be safe and our obligations to ensure this.

If you have any concerns about our products, you can contact us on ProductSafety@springernature.com

In case Publisher is established outside the EU, the EU authorized representative is:

Springer Nature Customer Service Center GmbH
Europaplatz 3
69115 Heidelberg, Germany

Batch number: 10620881

Printed by Printforce, the Netherlands